Jessica Wilson worked for South Africa's Council for Scientific and Industrial Research on integrated environmental management and environmental-impact assessments and has over ten years' experience in the area of environmental policy.

Stephen Law is the Director of the Environmental Monitoring Group, a South African policy and advocacy non-governmental organization.

A Brief Guide to
Global Warming

Jessica Wilson and Stephen Law

ROBINSON
London

The authors would like to thank Duncan Proudfoot, for thinking we could write this book, and for editing it; Harald Winkler, for pointing us in the right directions; and Victor Munnik, Liane Greeff, and Francis Wilson, for reading and encouraging us past our initial draft.

Constable & Robinson Ltd
3 The Lanchesters
162 Fulham Palace Road
London W6 9ER
www.constablerobinson.com

First published in the UK by Robinson,
an imprint of Constable & Robinson Ltd, 2007

A copy of the British Library Cataloguing in Publication Data is available from the British Library

ISBN-13: 978-1-84529-660-5

Printed and bound in the European Union

1 3 5 7 9 10 8 6 4 2

Contents

Introduction

There is sufficient, robust evidence that the world is warming and that the sea level is rising, and that these are due to human activity (in particular, the emission of greenhouse gases). This book will provide you with sufficient evidence and reasoning for you to hold your own in any argument. It also makes the point that it is as important how we solve the issue of global warming as that we solve it.

Is the world heading for the biggest disaster ever faced by modern humans? Is the world getting warmer and is this changing our climate? Is the modern industrial society engaged in a massive experiment with the world's climate, the outcome of which may see the end of civilization as we know it? It looks like it.

For some years, scientists recording global temperatures have noticed an almost imperceptible but steady rise in the average global temperature that had remained more or less constant for thousands of years. Glaciers are retreating, sea-ice is melting and plant and animal species are slowly moving towards the poles. Scientists have also seen a constant and rapid rise in the amount of carbon dioxide and other greenhouse gases in the atmosphere. There is now significantly more carbon dioxide in the air than there has been for the past 650,000 years (which is about as far back as we can measure). It seems that the gases we have been pumping into the air since the industrial revolution have had unintended consequences. What does it all mean?

This book explains why greenhouse gases matter, what their

impacts have been and are likely to be, and why it has taken us this long to wake up to the problem. Then we suggest what can be done about it. We hope that after reading this book you will be able to hold your own on any occasion at which views on global warming and climate change are exchanged in a loud, vigorous, and opinionated manner. We have tried to keep it simple. There are many other excellent books and websites if you want to know more.

The book has been divided somewhat artificially into "The Science" and "The Politics", though, of course, in the real world there is a symbiotic relationship between science and politics. We thought it best to start the book by putting forward the evidence that the planet is warming, in chapter one. Chapter two explains how this has come about – what the greenhouse effect is and why carbon is such a big deal. Chapters three and four both deal with the likely impact of global warming. Chapter three explains how climate models work, looks more closely at the climate system, and explains why such a seemingly small change in temperature can have such enormous consequences. The ice ages, for example, were sparked off by only a few degrees change in average global temperature. Chapter four explores how these climate changes in the physical world will affect us.

The second part of the book is essentially people's collective response to global warming. In this section, chapter five looks at how people are already adapting to a changing climate, and how to plan for future adaptation. However, we can only adapt to a certain extent and at a certain speed. What we really need to do is to stabilize and then reduce greenhouse gas concentrations in the atmosphere (this is examined in chapter six). Chapter seven looks at why we haven't yet done most of the things proposed in chapter six, which are no-brainers given the science involved. In chapter eight we pay a visit to the corridors of power and take a look at the political shenanigans that go on there. We see what governments have actually managed to agree to, and why it is not enough. Chapter nine suggests that the only truly fair way to tackle climate change would be to assign to each individual on the planet an identical quota of permissible greenhouse gas emissions from all activity (transport, heating, cooking, and so on) within a sustainable total. There could perhaps then be some mechanism for trading to increase one's individual quota, if others did not use theirs in full. It also describes some ways in which we can make the transition from the fossil-fuel-age.

Finally, in chapter ten we examine what we can do to help to prevent the planet from overheating.

So that's it. Enjoy the read. And remember to switch the light off when you go to sleep.

Section One

The Science

satellites could not be argued with. It is getting hotter and the rate of increase is frightening. The last century has been warmer than any other in the last 1,000 years. Of the twelve hottest years since 1850 when we first figured out how to measure temperature, eleven were between 1995 and 2006. The rate of warming over the last fifty years has been double the rate for the last 100 years. It is now, on average, 0.74°C warmer than it was a hundred years ago.

But hang on, you might say, what about all that rain and snow last winter? Well, let's be clear right from the start. We're not talking about the temperature in your back yard. We're talking about "average global temperature". That is the average temperature not only of every back yard in the entire world, from pole to pole, but also an average of daytime and nighttime temperatures and summer and winter temperatures. It is this average global temperature that is increasing.

But hold on again. If an average temperature is an average of all the ups and downs, then how can an average also go up (or down)? Well, it works something like this. Climatologists don't average out all the temperature readings ever recorded. That would give one number, the average, which would not be very useful. What climatologists are trying to uncover in the wild upswings and downswings of daily, monthly and annual temperature readings is a "temperature trend" (and for scientists, a trend has nothing to do with fashion – glasses and white lab coats are in again this summer). Imagine you're a scientist (Come on, try harder!) and that you are recording the temperature in your own back yard every two hours, every day, for a year and that you are plotting your results on a graph. The result would be a very jagged line, but through the confusing "noise" of daily fluctuations you should be able to see a general trend as temperatures fall in winter and rise in summer. Similarly, temperature records for the planet show a lot of "noise" from which a trend has to be determined. By using the right kind of statistical mathematics one can "smooth" the graph by eliminating the noise and allow the trend to be shown clearly.

Finding the trends hidden in the noise is one of the reasons it has taken us so long to start worrying about global warming and climate change. Past global temperatures, even those that have been averaged and smoothed, have had their ups and downs. In the last 1,000 years we have gone through significantly cooler and warmer periods. So the temperature rise that began in the early 1900s was not seen as anything abnormal. In fact for twenty years or so after 1945, it appeared that the world was actually cooling down again (sadly the effect was only temporary). It was only in the late 1970s that scientific consensus began to emerge, and even then, it was not a complete consensus and not too many scientists were alarmed. Exactly what it would take to alarm a scientist, we're not sure, but by the 1990s the alarm bells had started to go off in the ivory towers of academia. Even then, a handful of climatologists persisted in their belief that there was still insufficient evidence of global warming.

Was it a warmer world that was making the scientists and politicians so lethargic that they failed to recognize the looming problem? No, that's just silly. Well, maybe the politicians had dozed off, but the lack of alarm in scientific circles was in part due to the rigorous training that scientists go through. They are not allowed to believe anything until it has been proved a hundred times and has been published by an eminent scientist in a practically unreadable journal. The other problem is that there is no "provable" link between global warming and human-induced carbon dioxide emissions (there is more on this in chapter two). We have a Greenhouse Theory which, like Newton's Theory of Gravity, may not be perfect, but is as compelling as any of the existing generally accepted scientific theories. And we have a rise in greenhouse gas levels (of the kind which has not been seen for millions of years), which coincides with an exponential increase in our use of fossil fuels. Most non-scientists would put two and two together and head for the hills; but scientists are not like that. In 2001, the high church of global warming, the Intergovernmental Panel on Climate Change (IPCC), would only say that the observed warming was "likely" due to anthropogenic, or human-induced, emissions. Not exactly language to make people sit up and take notice, but they were getting there, slowly. In the 2007 pronouncement the language used was upgraded to "very likely". The next levels are "extremely likely", then "virtually certain", followed, perhaps, by "******* likely". (By the way, these terms all have statistical equivalents. "Likely" translates to 66 per cent, "very likely" to 90 per cent, "extremely likely" to 95 per cent and "virtually certain" to 99 per cent. As "*******" likely" is our own invention we can't claim that it has a statistical equivalent.)

The IPCC was set up in 1988 by the UN World Meteorological Organisation and UN Environmental Programme to help world governments make sense of global warming and climate change. Not only was there a confusing array of very technical information, but the scientists did not always agree with each other and delighted in poking holes in the data and conclusions of their colleagues. So the brief of the IPCC, a panel of a few hundred climate experts from around the world, was not to do any research itself but to consider the best and the latest research as well as commission work to fill any obvious gaps. They were then to write this up in a report that even a politician could understand, with all the appropriate "ifs", "buts", and "maybes". The IPCCs First Assessment Report was published in 1990 and by February 2007, the Fourth Assessment Report was undergoing final editing.

With each successive IPCC Assessment Report, the "ifs", "buts" and "maybes" have become fewer, and despite the authors of the reports covering their butts with careful scientific language, the message was loud and clear: global warming is happening and we are causing it. But how do we know?

One way to find out is to hop on a plane (on second thoughts, a rowing boat would be more climate friendly) and head for Iceland. About 10 per cent of the country is covered with glaciers with names unpronounceable to almost everyone except the Icelandic Glaciological Society. Each year for the past seventy years, the Society's volunteer members have been tramping out to the foot of the glaciers after the end of the summer melt, and measuring their positions. The Sólheimajökull glacier, for example, is about 300 metres shorter than it was a decade ago. The picture is much the same for all the others. Next-door Greenland is mostly covered by a huge glacier, so big it is called an ice-cap which holds in its frozen mass, more than 8 per cent of the world's fresh water. Just west of the very top of the ice-cap is a scientific station called Swiss Camp, built in 1990. If the ice-cap moves (as all glaciers do), the camp on its back moves too. Until 1996 it was moving at a rate of about 30 cm a day. In 2001 this had shot up to a lightning 45 cm per day. The sudden increase in speed was ascribed to melt-water making its way through the ice to the bedrock and acting as a lubricant between the ice-cap and the bedrock. Whatever the cause, more and more studies show that Greenland ice is indeed melting at unprecedented rates. Satellite probes show how the ice-cap is thinning – in some places by up to 15 metres since 1997. Even Glacier National Park in the USA is fast running out of glaciers and by 2030 may have to change its name or be sued for misleading advertising.

Some way south of the ice-caps but still in the chilly north, it used to be the case that one could dig down a metre or so and hit permanently frozen soil, called permafrost. However, in Alaska, since 1950 the temperature has increased by 3–4°C and begun to take the perma (or is it the frost?) out of the permafrost. The resulting melting causes the ground to shrink and split and crack, forming soggy open trenches beneath trees, roads, telephone poles and houses.

There is a lot more evidence for global warming in the northern hemisphere, partly because it is home to more scientific institutions and partly because (as the climate change models predict) the warming is more acute in the northern hemisphere because of its large landmasses. But the southern hemisphere is not without its climate change markers. The snow on top of Mount Kilimanjaro is disappearing at an alarming rate and may have all but disappeared in another twenty years while researchers still bicker as to whether this is a direct result of global warming or not. Much further south in the arid western parts of South Africa and Namibia grows a kind of tree-aloe called a Kokerboom or Quiver Tree. Its scientific name, *Aloe dicotoma*, illustrates the dilemma which this dichotomy posed to early taxonomists – is it a tree, or is it an aloe? This was nothing like the dilemma, however, that this poor aloe now finds itself facing. Researchers at the South African National Biodiversity Institute have noticed that over the last thirty years, aloe populations have been dying out rapidly in the northern-most extreme of its range. By comparing photographs shot from exactly the same location but decades apart, the researchers have also shown how the plants have grown well on the cooler mountain slopes, but died out on the hot valley floors. This plant, despite its exquisite adaptations to a hot, desert environment, is nevertheless being adversely affected by global warming.

The Kokerboom is not alone in its difficulties. Naturalists are becoming increasingly aware of some strange changes in the behaviour of some very specialized creatures. Take the insignificant Winter Moth, for example. Its even more insignificant larvae feed only on tender, newly-sprouted oak leaves. But while the oak senses the onset of spring by the lengthening of the days, the larvae get the signal to hatch from warming temperature. Over millions of years, the hungry caterpillars have finely tuned their hatching time to match the sprouting of new leaves. But recently things have gone awry. The little caterpillars are hatching earlier and earlier only to find that there are no leaves yet. Those who can hang on for a few days without food, and their sleepier siblings may survive. For the rest, it's a ticket to the great oak tree in the sky, courtesy of global warming.

Sifting through mountains of incredibly detailed records kept over the last century by nature-obsessed Victorian gentlefolk, ship's captains, amateur birdwatchers and others, researchers have built a detailed record of how around 1,700 different species are responding to global warming. Prior to 1950 there is little evidence of any pattern. But since then, there has been an average shift in habitat towards the poles by about 6 km per decade and up mountainsides by about 5 metres per decade. These moves towards the poles and to higher, cooler altitudes are hard to explain, except in terms of global warming. Not every plant and animal is moving to cooler climes, though. Some plants are starting to bud or flower a few days earlier, and migratory birds and butterflies are arriving a few days earlier.

And then there is *Bufo periglenes*, the Golden Toad. Some suggest that this is the first observed species to have been made extinct as a result of climate change. Biologists had only discovered the Golden Toad on the high slopes of Costa Rica's mist-shrouded mountain rainforest in 1966, and by 1987 the species was extinct. The clouds and mists that used to keep the

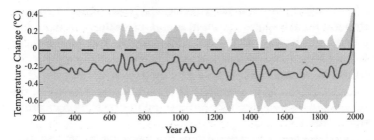

Departures in global surface temperature in °C from the 1961–1990 average
between AD 200 and AD 2000.

rainforest and frogs supplied with moisture throughout the dry
summer were also gone. A sudden rise in ocean temperature
had lifted them up to only the very tops of the mountain peaks.
Rest in peace, Golden Toad.

Scientists say that species are disappearing and glaciers are
retreating because the world is warming. Do the thermometers
confirm that the temperature is increasing? Yes, they do. Since
the 1850s when detailed record-keeping began, the world has
grown warmer by almost 0.8°C. And scientists can tell us even
more. They have used all kinds of techniques to estimate the
temperature of the earth for the past two thousand years. This
information is captured in the now famous "hockey-stick"
graph.

The graph, which plots estimates for average global temperature between the year AD 200 and the present, looks like an ice-
hockey stick on its side. The long flat handle represents the
relatively stable temperature that the world has enjoyed for
almost 2,000 years. On the right-hand end of the graph, the
short, steeply angled section gives an indication of exactly how
fast things are changing.

Since recorded temperatures stretch back only to the relatively
recent past, scientists have had to make use of "proxy" data,
typically information from tree rings, ice-cores, corals, lake and
sea sediments and the like. Tree rings indicate how fast a tree
has grown in a season. Ice-cores record snowfalls. Sediment-
cores capture dust, pollen, plant and animal remains and other
detritus. All of these allow one – with the right scientific creden-

tials and methods – to make an educated guess of the prevailing temperature at a specific point in time. One of the first to publish such a graph was American climatologist and geophysicist Michael Mann and colleagues, in 1999. Of course scientists are trained to be sceptical, and seldom believe what they are told, and it wasn't long before poor Dr Mann found himself in the middle of a vicious bunfight around the interpretation of proxy data.

This has not detracted from the consensus of scientists throughout the world that the planet is warming. The IPCC, known for its measured statements, assure us that warming of the climate system is *unequivocal* – it is happening. But *why* is it happening? What is causing the temperature to rise, the ice caps to melt and species to adapt or die? Chapter two will address these questions.

The greenhouse effect

Way back in the 1820s a French mathematician and physicist named Jean Baptiste Joseph Fourier thought about the temperature of the earth. Too warm, he thought. The earth should be losing as much energy into space as it receives from the sun and theoretically should therefore be colder than it is. The atmosphere, he hypothesized, was acting like a gigantic greenhouse roof and trapping heat – sunlight could come in, but heat could not get out. This is good since without the greenhouse effect the earth would be terribly cold (minus 18°C to be precise).

Taking Fourier's hypothesis seriously almost forty years later, John Tyndall, an Irish-born scientist, measured how much heat the atmospheric gases absorb. Water vapour and carbon dioxide absorb heat, he concluded; oxygen and nitrogen don't. Well, that was helpful and borne out by what we know about the other planets. Mars with no carbon dioxide is very cold whereas Venus with an atmosphere of 96 per cent carbon dioxide is as hot as hell at 470°C. Since nitrogen and oxygen together account for 99 per cent of the gases in the earth's atmosphere, we don't have to worry about most of it. In fact the balance is rather extraordinary – just enough oxygen to keep us alive without spontaneously combusting, and the minuscule amount of carbon dioxide present is just the right amount to keep us comfortably warm, neither freezing nor boiling. But this minuscule amount is increasing, which is a cause for concern. Here is how it works.

The sun is a star and pours out energy generated from nuclear reactions going on in its core. It is a furnace with an almost never-ending source of fuel. (In fact, the sun will one day burn itself out, but this will happen so far into the future that it should not worry even the most neurotic among us.) Sunshine enters the earth's atmosphere and, depending on its wavelength and what it encounters on the way, is either reflected back into space or continues its journey towards the earth. A portion of the sun's energy is absorbed before reaching the earth. For example, ultraviolet light hitting an ozone molecule (three atoms of oxygen linked together) high up in the stratosphere will be absorbed by the ozone and will not reach the earth. This is one of the reasons we are so worried about the hole in the ozone layer that lets through UV rays and gives us skin cancer, but that's a story we'll tackle a bit later on when we look at the politics of tackling environmental problems in chapter eight. (It is also worth noting that ozone is a pollutant we produce that acts as a greenhouse gas, but only when it is found close to the earth's surface.) The energy from the sun, primarily in the form of UV and visible light, that does make it all the way through the atmosphere and to the surface of the earth is soaked up by the sea and land (think how lovely and warm rocks can be after a day of sunshine). This heat is radiated back into the atmosphere. While greenhouse gases are transparent to light (they let it through), they are opaque to heat (they reflect it back). On the whole greenhouse gases have made our planet a very pleasant place for human beings to live.

The higher the concentration of greenhouse gases in the air, the more heat is trapped. There are several greenhouse gases, but to keep things simple they can all be compared to carbon dioxide and are given a carbon dioxide "equivalence". From a global warming perspective, carbon dioxide is the one of most concern to us. To understand why, we need to take a brief detour into the history and behaviour of carbon, which is the history of life on earth.

The carbon cycle

Carbon is everywhere. It is found in the oceans, air, soil, plants, and animals. It is a multi-purpose element able to bond with many other elements to form solids, liquids, and gases. In its pure form it can bond with itself in different ways to form substances as different as soft, slippery graphite or harder-than-rock diamonds. This ability to take on different forms and bond easily with itself and many other elements is one of the reasons why carbon is so central to all living creatures. Its various compounds make up the protein in our DNA, our bone and muscle and other tissue, and the fatty layer under our skins. It also makes up most of the food we eat and carries in its chemical bonds the energy we need to stay alive.

Like water which evaporates from the sea to fall as rain, so carbon compounds are continuously being recycled. Carbon is our living planet's complex breathing and energy-exchange system.

The energy contained in carbon compounds comes not from the carbon itself, but from its chemical bonds with other elements, most notably hydrogen, with which it forms carbohydrates, like starch and sugar.

Plants don't need much energy. They get what they need from the sun, but put it to spectacular use. Through tiny pores in their leaves, they draw in carbon dioxide (CO_2, a carbon atom with two oxygen atoms attached) from the air. With chlorophyll as a catalyst, this carbon dioxide is combined with hydrogen in the water (H_2O, two hydrogen atoms each connected to an oxygen atom) brought up by the roots. The sun provides the energy to rearrange the molecular bonds which results in the creation of a simple, sugary carbohydrate and a bit of "waste" oxygen (O_2), which finds its way out of the leaf and into the air. The newly formed carbohydrate molecule is used by the plant as a carbon and energy source to build roots, branches and new leaves.

Animals, on the other hand, do require a lot of energy. So, from the smallest bacteria upwards, they eat plants or other animals that have eaten plants, or other animals that have eaten animals. The food chain is an endless cycle of dog eat dog, with photo-synthesising plants at the very bottom of the pile. Instead of carbon dioxide, animals breathe in oxygen (O_2) which is used in their cells to break the carbon-hydrogen bonds of the hydrocarbons (C_xH_y) that the plants have so carefully assembled. When these bonds are broken, energy is released along with the two waste products, carbon dioxide gas (CO_2) and water (H_2O).

Animals depend on plants for carbohydrates. Plants in turn need the waste carbon dioxide from animals. This gas, contained in the air in tiny quantities, is key to this delicate and beautiful relationship. A slightly smellier exchange takes place with organisms that don't breathe oxygen. Here other mechanisms are used to break down hydrocarbons and the resulting waste product is methane (CH_4), which is also a greenhouse gas.

The movement of carbon through the planet's living system, the carbon cycle, involves not just carbon in the air, but also the carbon stored in trees in the great forests, in the wide grasslands,

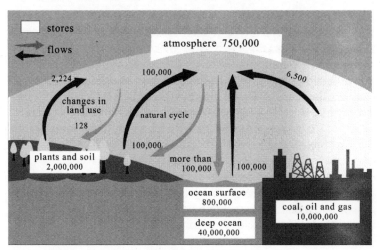

Carbon cycle. Carbon stores and annual flows in million tonnes.

in the bogs of the tundra and in the organic matter in the soil below our feet. The greatest carbon storehouse of all is the sea where huge amounts of carbon dioxide gas are dissolved. Tiny and not so tiny sea creatures use dissolved carbon dioxide to help build their calcium carbonate shells and skeletons. Parts of the cycle where carbon is "captured" and stored for more than just a short period are known as "carbon sinks", the opposite of "carbon sources".

The mechanism governing carbon accumulation and circulation between the various parts of the system and the rate at which this occurs has evolved over aeons, and has been fine-tuned over the course of the last few million years to exactly the right balance by living organisms themselves.

A few million years ago, the earth was a very different place. People had not yet evolved; large parts of the planet were covered in swampy, shallow lakes; there was lots of carbon dioxide in the air and the climate was warm and humid. As a result, plants grew like crazy. Ferns, mosses, primitive trees and giant insects ruled. Swamp-loving bacteria were not far behind. They soaked up masses of carbon dioxide and, then, like all living things, they died. Some did not decompose and release their carbon completely, but sank to the bottom of the bogs and swamps. Over millions of years, the remains of these partially decomposed plants and bugs were covered over with earth and compressed. The earth's geological movements pushed them deep under the surface where they became fossilised. And, hey presto! We have coal. Some swamp bottoms remained in an oily, liquid form, which we know as, you guessed it, oil. Some ended up as something in between: a kind of squishy tar, or perhaps an oily grit.

Even underwater animals accumulated and stored carbon. Molluscs and plankton extracted carbon dioxide dissolved in water and used it to build their little shells and skeletons by converting it into calcium carbonate ($CaCO_3$). In the same way that coal formed, these shells and skeletons became fossilised into what we know as limestone, a key ingredient in the manufacture of cement.

The earth's carbon balance has been relatively stable for millions of years. Then we discovered ways to burn coal and oil in vast quantities and to make cement. We released carbon dioxide that had been stored for aeons, locked out of the carbon cycle, and, as a result, the world started to warm up.

The industrial revolution and the burning of fossil fuels

If you had taken a sample of air in 1769 when James Watt patented his coal-fired steam engine, you would have found that approximately 280 parts out of 1,000,000 parts (280 ppm or parts per million) were carbon dioxide. This is only 0.028 per cent or less than three parts per ten thousand. It sounds so very little. Nevertheless, it is extremely significant and was about to grow. And this – though no one knew it at the time – was cause for alarm. Coal had been used in Britain since the bronze age four to five thousand years ago, but in the eighteenth century coal mines were in danger of closing because it was too expensive to drain them of water. Watt's invention changed all that. The mines could be drained more cheaply and the coal could feed steam engines to do all kinds of other things. The industrial revolution, heralded by the coal-age and steam engine, was the start of human-induced climate change.

In 1896 a heavyweight Swedish scientist called Svante Arrhenius, weighing over 200 pounds, thought about emissions from industrial processes. When coal – or any carbon-based fuel such as oil, wood, or the now popular ethanol – burns, it combines with oxygen to produce energy, carbon dioxide and some nasty pollutants. Watt's steam engine was no exception. Burning coal produced energy that heated the water to create the steam to turn the pistons, and carbon dioxide, an odourless, colourless, non-toxic gas was emitted into the atmosphere, along with a number of other noxious substances. Remembering Tyndall's discovery that carbon dioxide absorbs heat, Arrhenius did some calculations to estimate what the temperature of the earth would be if carbon dioxide from industry caused concentrations to double. His estimate of 5 to 6°C compares with today's best estimates of 2.7 to 4.3°C. He was, however, very wrong in his prediction of how long it would take to double atmospheric carbon dioxide. His estimate was 3,000 years; today's estimate is the end of this century, unless we take drastic action now.

You know that a fuel is carbon-based and will produce carbon dioxide if, at some point in its life, it was once a living being or part of a living being. In the case of wood and charcoal it is obvious that they come from trees. What of ethanol? This is usually produced by fermenting sugar cane or another agricultural crop or waste product. These are often called renewable fuels because it is possible to burn a tree while growing another one in its place. Burning a tree releases the same amount of carbon dioxide as it would have done in the natural process of decomposition, though at a slightly faster pace. For this reason, growing and burning these so-called biofuels does little either to add to or to subtract from the build-up of carbon dioxide in the atmosphere. (There is more about biofuels and renewable energy as alternatives to coal and oil in chapter six). Fossil fuels are also carbon-based, having once been alive as plants or animals.

Just as burning fossil fuels releases carbon dioxide, the process of turning limestone into cement also releases carbon dioxide as a waste product. To get calcium oxide (CaO), the key ingredient in cement, limestone (calcium carbonate or

$CaCO_3$) is heated and carbon dioxide (CO_2) is released as a by-product. (Readers who are chemical engineers will know that there are also silicates involved, but these are not important from a global warming perspective). Making cement is a greenhouse double whammy because fossil fuels are used to heat the limestone. For every tonne of cement produced about half a tonne of carbon dioxide is released *excluding* the carbon dioxide released from energy needed in the process. In total, close to a tonne of carbon dioxide is released for every tonne of cement made!

Whether hard or soft, oil or limestone, fossil deposits represent millions of years of hard work by plants and animals using sunlight to capture and extract carbon from the atmosphere. When we burn fossil fuels or convert limestone to cement, we are returning to the air all the carbon that was extracted millions of years ago at a time when humans did not exist. It is not surprising that we are disrupting the carbon cycle.

No one thought to measure how much carbon dioxide we were adding to the air until 1958 when an American scientist called Charles David Keeling decided to make some accurate measurements. This began for him almost as a hobby because he loved the outdoors. He spent a year building his own carbon dioxide measuring apparatus, which he used to take on camping trips with his wife and small child. After taking readings at campsites across the USA he eventually got permission to set up a measuring station at Mauna Loa observatory in Hawaii where he measured an astonishing 317 parts per million of carbon dioxide, rather more than the pre-industrial levels of approximately 280 ppm. Then the level started to go down (you can imagine the relief and smug, I-told-you-so smiles of the oil barons of the day). But any relief was short-lived; at the end of the northern hemisphere summer, the concentration of carbon dioxide in the atmosphere started to go up again. This cycle is captured in a now famous graph, the jagged "teeth" of which represent the seasonal breathing of plants in the northern hemisphere.

The northern hemisphere "breathes" more than the southern hemisphere because it contains much more land, particularly at high latitudes, than in the south. Plants at high latitudes lose

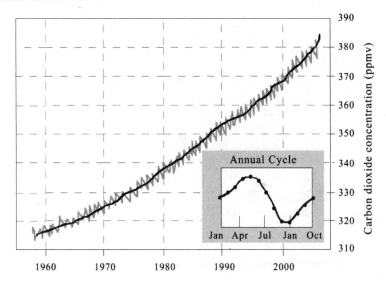

Keeling Curve: atmospheric carbon dioxide measured at Mauna Loa, Hawaii.

their leaves and generally don't grow as much in winter as they do in spring. Keeling's jagged-tooth graph does not lie horizontally, but climbs over time showing clearly that although there are seasonal changes, the amount of carbon dioxide in the air is rising steadily. By 2005 it had reached 379 ppm.

While Keeling's own personal carbon is now being returned to the earth his name lives on and the Mauna Loa measurements continue. Not only do they reflect the carbon dioxide emitted from burning fossil fuel, they also show the slow release of carbon from deforested land. Remember that forests store a lot of carbon and chopping them down releases their stored carbon into the air through decomposition (being eaten by bugs and fungi) or burning. Even trees that are converted to paper or other products will release their carbon when the product decomposes in a landfill site.

Between 2000 to 2005 we poured around 26.4 gigatonnes of carbon dioxide ($GtCO_2$) *every year* into the air from burning fossil fuel and making cement. (Some people measure emissions in

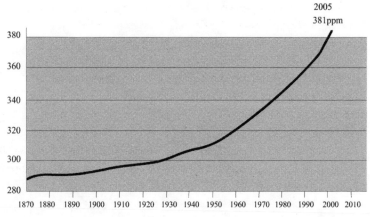

CO_2 in the atmosphere. Global concentration 1870–2005 in parts per million.

terms of carbon not carbon dioxide. To get the carbon equivalent, divide the weight of carbon dioxide by 3.667). A tonne is a thousand kilograms; a gigatonne is a billion (1,000,000,000) tonnes. To give some idea of what this means, consider a male African elephant, the biggest mammal on earth with a weight of between five-and-a-half and six tonnes. Every year we pump into the air carbon dioxide equivalent in weight to four-and-a-half billion African elephants. Carbon dioxide emissions from changes in land use are much harder to estimate, but they were around $5.9GtCO_2$ per year in the 1990s, or over a billion African elephants.

What on earth do we do with all the fossil fuel we burn, the cement we make and the land we clear of forests? It's probably obvious. We heat, build, and light our homes and grow our food. But we also fly in very fast aeroplanes to visit the last of the coral reefs, manufacture steel to build tanks and bombs, and spend a lot of time alone in our cars stuck in endless traffic jams cultivating road rage. We have in fact built our entire economies, the wealth of our wealthy nations on burning fossils. Fossil-fuel energy is embedded in pretty much everything we use or consume. This dependence is fairly recent, dating only since industrialization – a mere moment in the history of our species. (This will be discussed further in chapters six and seven.)

Methane, nitrous oxide and three synthetic gases

Remember Tyndall and his experiments way back in the 1860s? He found that carbon dioxide was not the only gas to absorb heat. Water vapour does, too. And so, we now know, do methane (sometimes called natural gas), nitrous oxide (from fertilizers and the partial combustion of coal and oil), sulphur hexafluoride (used as an insulator for circuit breakers), chlorofluorocarbons (those terrible synthetic chemicals responsible for eating up the ozone layer), other halocarbons (HFCs, PFCs), and ozone itself when in the lower atmosphere or troposphere as opposed to the stratosphere where it protects us from UV. Aside from water vapour, which needs to be treated as a special case, human beings produce all these and all of them contribute to global warming. The greenhouse gases differ in their potency (how much a molecule of each warms the earth) and how long they stay in the air before being broken down or absorbed. To simplify things for us, the scientists and policy-makers have figured out a way of comparing them all to carbon dioxide, because carbon dioxide is the most significant. This is called their carbon dioxide equivalent or CO_2e. (There are also emissions arising from human activities that have a cooling effect on the planet, sometimes referred to as global dimming. The magnitude of these is far smaller than that of the warming gases, nevertheless they are discussed in chapter three.)

The most significant greenhouse gas resulting from human activity after carbon dioxide is methane, which actually has a higher global warming potential per molecule than carbon dioxide. Averaged over a hundred years, each kilogram of methane warms the earth twenty-three times more than each kilogram of carbon dioxide. This is known as its global warming potential (GWP). There are some greenhouse gases with even higher GWP values than methane such as sulphur hexafluoride, which has an extraordinary GWP of 22,200 over 100 years, but luckily these occur in very small quantities. It is lucky, too, that methane is short-lived in the atmosphere (breaking down into carbon dioxide and water vapour fairly rapidly). Methane is also present in smaller quantities than carbon dioxide so its overall effect is not as great. Nevertheless it contributes significantly to global warming. While we have increased carbon dioxide concentration by a third, we have already more than doubled methane concentrations. Before the industrial revolution, there were 715 parts per *billion* (ppb) of methane; now there are almost two parts per *million* (to be precise 1.774 ppm). The level of methane has risen particularly rapidly over the last hundred years. Apart from being released (or escaping) as a result of mining and the processing of fossil fuels, it also forms as a result of the activity of anaerobic organisms – tiny bugs that thrive in oxygen-free environments like our gut, waste dumps, and the bottoms of large dams and rice paddies. Interestingly, cows raised in intensive feedlots contribute a lot of methane through farting, while their freely roaming grass-eating relatives in India contribute very little.

The link between increased greenhouse gas concentrations and rising temperatures is disputed only by loony scientists and flat-earthers. The rest of the world is convinced of the connection. So let's continue by considering just what a warming world will look like.

Chapter Three

Weather, Climate, Models and Ice Ages

Temperature, particularly a change in temperature, is a key driver of climate. Climate is a complex system with many variables that create both positive and negative feedback loops. People have learnt how to model climate, which helps us predict the future, which, in turn, helps us to plan or to take precautions. Models are based on probability and likely scenarios. Since there is more energy in the climate or weather system, weather events are more extreme as well as generally warmer.

With all the hype around global warming and climate change, you might think that this has never happened before. You would be wrong. Fossil and ice-core evidence shows that the earth has gone through many, regular climate changes, and that these have been quite dramatic. In the last 20 million years or so the planet has experienced regular ice ages which have lasted around 100,000 years, interspersed with warm interglacial periods of between 8,000 and 40,000 years. The last ice age ended about 18,000 years ago, so we are due for another some time soon. The swings between ice ages and interglacial periods have been extremely dramatic. Yet they were caused by a global average temperature change of only 2 or 3°C.

In the 1920s, Serbian mathematician Milutin Milancovitch postulated that predictable "wobbles" in the rotation of the earth over tens of thousands of years, which would change our angle of orientation to the sun and therefore change the amount of solar radiation received by the high latitudes, could be the cause of ice ages. By the 1980s, analysis of ocean sediments had confirmed this correlation between the rotational wobbles and the cycle of warm periods and ice ages, now known as Milancovitch cycles. But because of the extreme changes, there must be other powerful positive feedback effects that kick in and make the warm periods warmer and the cold periods colder.

Overlaying this long-term and relatively steady pattern is the more erratic appearance of sunspots which may influence the amount of solar radiation received. Another source of heat, the internal heat of the planet, though massive, is a constant and does not play a significant role in planetary warming or cooling. As we will see, neither of these can compete with the massive effect of greenhouse gases.

Ice records show that there is a strong correlation between temperature and carbon dioxide concentrations. When one goes up, so does the other, and likewise when one declines. In the past, carbon dioxide has lagged behind temperature which implies that a warmer world releases more stored carbon into the atmosphere. Now, carbon dioxide is leading as a result of human emissions. The effect of a warmer world on stored carbon dioxide is yet to be seen.

This is what this chapter will explore, as well as showing how temperature – which is increasing as a result of all the green-house gases that we are pumping into the air – is a key driver of climate. To do that we need to be clear about what climate is and how it differs from weather.

Climate and weather
We all know that the weather is constantly changing. Some days it rains and some days the sun shines. Who knows? Tomorrow

might be warm and still, or windy and cool. That's the weather, always changing. Climate is different from weather, though. It doesn't change from day to day – or not yet, anyway. We talk about a place having a particular climate, which is a kind of summary of the average weather characteristics of that place. The south pole, for example, has an icy cold climate with blinding snowstorms which isn't much fun, unless you're a penguin, and probably not even then, really. The Caribbean, on the other hand, has a great climate by all accounts. It's warm there and the wind hardly ever blows, apart from during the hurricane season. It would be a fun place to live, with al fresco bars on just about every beach and the biggest danger, falling coconuts.

Climate is more or less the equivalent of the "average" weather for a particular place. But the average weather can also change. There are hot years, cold years, floods, and seven-year droughts. Sometimes they even manage to play some tennis at Wimbledon before the rain comes down. So to accurately define a particular climate and avoid seasonal and other anomalies, the weather needs to be averaged out over a very long time period. You would imagine that old people would be good at this, having lived through many summers and winters, but they're not. Some very old people with good memories may recall the terrible heatwave of '24, for example, which they toughed out without air-conditioning, but not a great deal else about the weather that was the backdrop to their lives. Fortunately there is a dedicated band of scientists called meteorologists who make it their business to collect and analyze information about the weather and they have amassed heaps of data from weather stations around the world. These scientists have an international club called the World Meteorological Organization, where they drink beer and talk about the weather – probably not all that different to your local pub or club. These meteorologists have suggested that averaging weather over a period of thirty years is sufficient to define climate.

So what makes a climate what it is and what might change it? Well, mostly it has to do with the energy that moves wind and ocean currents about, makes it rain or snow, and so on.

Since global warming means higher temperatures and more energy in the climate system, it should be obvious why the terms global warming and climate change are used almost inter-changeably. Still, the interactions within the climate system are complex and have spawned a whole bunch of people called modellers who attempt to figure out the interactions with enough precision to predict how climate will change. How do they do this and what do they include in their models?

Predicting the weather

In the good old days, no one bothered with weather or climate models. The Germans used to say, "When the rooster crows on the dungheap, the weather will change or stay as it is". Empedocles, an ancient Greek, said that weather was caused by a competion for dominance between the four elements of earth, air, fire, and water, but he neglected to explain how. Later, in the eighth century, the Venerable Bede postulated that clouds caused wind. We know a five-year-old who postulates – based on direct observation – that wind is caused by trees waving their branches.

Norwegian physicist Vilhelm Bjerknes was the first to propose in the 1890s that there might be a correlation between weather patterns and mathematical and physical laws. If this was so, he said, it would allow the weather to be predicted mathematically – but he didn't drive himself mad actually trying to do it. This was left to an Englishman, Lewis Richardson, who was already half-mad, or, at the very least, extremely eccentric. As an ambulance-driver during the First World War, he developed his seven complex weather equations during lulls in the fighting. He lost all his papers in the chaos of war, but luckily found them again in Belgium under a pile of coal. His papers were expanded into a publication, but ironically the coal was burned. In his ground-breaking book, Richardson spelled out his mathematical theory of weather. He also envisaged the world's first weather/climate-predicting computer.

He dreamed of a huge hall with a map of the world painted on the walls. Arranged along each side of the enormous room were galleries packed with men wielding state-of-the-art slide-rules and pencils, each one working to solve a small part of an equation for a small part of the earth. Answers were passed on to the next person to continue the calculation. In the middle of the hall, in a sort of pulpit, was a man waving red, green, and blue lights, who was conducting the whole operation. Richardson calculated that he would need only 64,000 people to predict the weather as fast as it was actually happening! To make any useful future predictions would need many times that number. Much later, electronic computers would allow for lightning-fast mathematically-based predictions and do away with 64 000 potential jobs. Although nowadays even the most powerful computer still takes about a month to complete a hundred-year prediction. Why? What makes it so complex? We will attempt to explain how sunlight, oceans, air, clouds, earth, and mountains interact to produce "climate" and why our meddling with a tiny part of the composition of the air's gases makes such a big difference.

It's all about energy

One of the first things a modeller does when he or she sits down to work is to figure out how much *energy* there is in the system. This is not an easy thing to do in your spare time, but after reading this section you should at least know what numbers and factors are fed into the computer. Remember that energy, in the form of temperature, is a major driver of climate.

Energy? Temperature? It should be obvious that the sun plays a major role! There may be more or less sunshine, but the sun is always lurking in the background, a kind of benevolent, cosmic big brother. We call it "sunshine" – climatologists call it "incoming solar radiation", another scientific expression with which to impress your friends. So, for a start, the climate of a particular region depends on how much sunshine it gets and how strong that sunshine is. The strength of the sunshine and the amount of warmth it brings is largely a factor of the angle between the earth's surface and the sun. Does this sound too complicated? Imagine sitting at the south pole on a clear day at the height of summer. After having accustomed yourself to a cold bottom, you would notice that the sun, even though it shines all day and night at the poles during the summer, nevertheless barely clears the horizon. It remains at a very low angle to the surface of the earth and doesn't provide much warmth. At the equator, on the other hand, although the sun only shines for twelve out of twenty-four

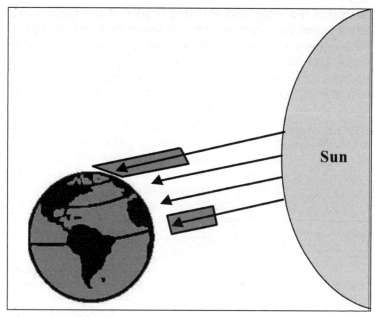

The strength of the sunshine and the amount of warmth it brings is largely a factor of the angle between the earth's surface and the sun.

hours, at noon it is directly overhead, forcing all but mad dogs and Englishmen into the shade. It is all to do with the angle of the sun to the earth. The closer to 90 degrees the angle, the more energy the earth receives. The poles and the tropics are the two extremes; in the temperate regions the sun is at an in-between angle, and the climate is, well, temperate. So the first general rule of climate is that the further away from the equator one goes, the colder it is. Those bright modellers would say that climate is a function of (i.e., depends upon) *latitude*. But it depends on more than just latitude, as you will see.

Irrespective of the angle of the sun there are other things – like clouds, umbrellas, and air pollution – that affect the amount of sunshine that manages to reach the earth. Modellers need to take some of these into account and they do so by estimating their "radiative forcing". This term isn't as complex as it sounds; it is simply a measure of how much a particular factor alters the balance of incoming and outgoing energy. If the factor has a "positive forcing" it means it tends to warm the earth's surface; if it has a "negative forcing" it means it tends to cool it. On the whole, air pollution has a negative radiative forcing. In simple terms it blocks the sunlight. But let's look at that in a little more detail.

Little particles of soot, ash, and dust are continuously being wafted upwards by smokestacks, vehicle exhausts, and so on. These, together with windblown dust and volcanic emissions, are called "particulates", and they can block out significant amounts of sunlight. Burning oil and coal also releases oxides of nitrogen and sulphur (commonly known as NO_x and SO_x) which rise high into the atmosphere, and combine with water vapour and particulates in complicated ways which we won't go into here, to form tiny suspended droplets called "aerosols". The important result is that significant amounts of sunlight may be prevented from reaching the earth by these particulates and aerosols, causing a drop in global temperature. This is also known as "global dimming". This process is bad for your chest, but good for preventing global warming. The effect that aerosols have on global temperature was seen after 1945. The upward temperature trend at that time actually reversed and the world started cooling. By the 1960s, however, air pollution had been found to be a major cause of the acid rain that was killing forests and dissolving marble statues, and by the end of that decade, most industrialized countries had moved to substantially clean up emissions from their smokestacks. So successful was this clean-up, in fact, that the world resumed its warming trend. The global warming debate has tended to lure large numbers of crackpots out of the woodwork, and, unsurprisingly perhaps, some have suggested that we may be able to reverse global warming by increasing the quantity of pollutants that we pump into the air.

You have probably realized that greenhouse gases contribute to *positive* radiative forcing (i.e., they warm the earth's surface), but by how much? What's the big deal? Well, the last time the world's scientists got together to agree on what they could agree on, they estimated that the amount of carbon dioxide we have poured into the atmosphere since 1750 has a positive radiative forcing of 1.66 Watts per square metre. You could think of that figure like this: if you could rig up a system of one 1.66-Watt lightbulb on every square metre of the earth's surface, that is how much warming would be due to anthropogenic (of human cause) carbon dioxide. The power of a standard household lightbulb is 60W, so this would be equivalent of placing one 60W lightbulb every 36 square metres over the entire surface of the planet. Methane has a radiative forcing of $0.48W/m^2$ (just over a quarter of that of carbon dioxide), nitrous oxide of $0.16W/m^2$, and halocarbons (which include CFCs, HFCs, and PFCs) of $0.34W/m^2$. These might just look like numbers, but they are important when calculating the full impact of all greenhouse gas emissions. And luckily there are scientists who are skilled at building them into climate models.

Another factor modellers include is how much sunlight is reflected back into space. The polar regions are cold not only because they are so far north or south but because there is also a feedback double-whammy effect (not a proper scientific term). Ice is a very good reflector of sunlight, and a lot of sunlight gets reflected back into space (as well as into eyes, causing snow-blindness) before it has a chance to warm anything up. If global warming causes some of the polar ice to melt, then less sunlight is reflected, and more is *absorbed* because water absorbs the heat from the sun very effectively. There is, in fact, a term for how much sunlight a particular material absorbs or reflects: the "albedo effect", which is expressed as a number between 1 and 0 where 1 indicates total reflection and 0 indicates total absorption. On average, the earth has an albedo of 0.3 meaning that almost a third of the sunlight which reaches it is reflected back into space. Ice has an albedo number between 0.8 and 0.9, whereas the albedo of the ocean is around 0.07, meaning that less than 10 per cent of light is reflected back into space. Needless to say, modellers build these numbers into their models, which in turn illustrate the magnifier effect of turning ice – one of the best *reflectors* of sunlight – into water – one of the best *absorbers* of sunlight. So, when the earth's temperature rises (due to those pesky greenhouse gases) ice melts, causing less sunlight to be reflected which makes things even warmer, and the warmer it is, the more ice melts, and the less ice there is, and the less sunlight is reflected and the warmer it gets . . . Do we need to go on? When things get colder, this can happen in reverse too.

The important thing is that such positive feedback effects can turn a small change in temperature into a big change in climate. It is like a tiny key being able to unlock a huge door, and it is thought that such feedback effects (for there are others, too) are the reason why certain dramatic climate changes happen rapidly, over decades, while others take place over thousands of years.

It is widely accepted that past ice ages have been strongly driven by this kind of "runaway" feedback effect. There is even a double-whammy inside this double-whammy. If temperatures rise to the point where there is large-scale melting of permafrost (permanently frozen earth), millions of years of dead and as yet undecayed vegetation will begin to rot. Apart from the smell and the swarms of bog flies, billions more tonnes of carbon dioxide and methane will be released into the atmosphere, adding to the already high levels, and warming the planet further. It is estimated that if all the methane from the West Siberian bog was released, it would be the equivalent of adding another seventy years of human-generated carbon dioxide at current rates.

What about clouds, oceans, and other wet things?
No one can talk about the weather without mentioning rain, so modellers need to be on top of this, too. In many parts of northern Europe people seem resigned to one rainy day after another. On the other hand, in many parts of southern Africa rain is a symbol of renewal and a blessing. There is an intimate connection between all things watery – rain (or drought), storms, clouds, and oceans . . . and, of course, temperature. Let's look first at the sea because over 97 per cent of all water on the planet is found there.

The oceans are important determinants of climate because water has a high "heat capacity". This means it can store heat well – and a lot of it. Boil a kettle of water and note how long it stays warm. Now do the same with air – you may end up ruining your kettle, but you'll be doing some real science! Notice how much easier it is to heat water and keep it warm than it is to do the same to air. The oceans soak up a huge amount of global warming, an estimated 80 per cent, and because the oceans are deep and constantly mixing, temperature changes in coastal areas may be less pronounced. For the same reason, temperature in the southern hemisphere, which contains most of the world's oceans may rise more slowly. The oceans also introduce a lag effect so that even if we stabilize greenhouse gas concentrations today, global warming would continue and the sea level would continue to rise for more than a millennium.

Warmer oceans also mean expanded oceans, which means a rise in sea level. Heat causes seawater to expand just that tiny bit. Multiplied across all the litres of seawater this amounts to a huge increase. How high and fast the rise will be depends on many factors, but the model scenarios predict that it could be as much as a 59-cm rise by the year 2100. This may not sound like much, but remember that this is just an average and does not reflect local extremes.

The rise in sea level could be significantly greater. If the Greenland and Antarctic ice-caps melt, they will release trillions of tonnes of water that is now "locked up" as snow and ice above sea-level. If the Greenland ice sheet melted completely it would add 7 metres to the rise in sea level. For now, the IPCC predict that contraction of the Greenland ice sheet will continue but that the Antarctic ice sheet will thicken due to increased snow. The dynamics of these processes are not yet well understood or agreed on by scientists. (Some predictions see an almost total disappearance of Arctic sea-ice by the end of the twenty-first century. Although this is a terrible prospect for many reasons – consider the polar bears – melting ice-bergs do not add to the rise in sea level because they displace as much water as they contain).

Oceans have the capacity to store a lot of heat; they also move it around the world like a kind of huge thermal conveyor belt, and the force that drives these currents is . . . temperature. For example, cold water in the polar regions sinks (because it is dense) to be replaced by warmer water from the tropics. This circulation warms the polar regions and cools the tropics. Scientists think that there may be powerful feedback effects at work in the way oceans currents work, particularly their role in melting polar ice. They fear that global warming may disrupt currents and lead to rapid and extreme changes, or so called "flip-flops" in the earth's climate. Ice-core evidence from Greenland and Iceland shows that climate change is not necessarily the slow and gradual change over centuries that models predict. For reasons not clearly understood, the earth's climate has seen a number of very rapid and runaway changes in our recent (geologically speaking) past – changes which have taken place over decades. So if we interfere with the global temperature, we interfere with currents, with dramatic and unpredictable effects.

The Gulf Stream illustrates the role of ocean currents in climate. Glasgow and Moscow are on practically the same latitude. But while fur-clad Muscovites drink vodka and sing about death and revolution in winter temperatures of minus 20°C, Glaswegians seldom experience winter temperatures much below minus 5°C and are able to wear nothing beneath their kilts, even in the middle of winter. This is all thanks to the Gulf Stream. This ocean current originates in the warm Gulf of Mexico where it picks up masses of energy, and, lured by opportunities in Europe, flows northwards at a rate of knots (4, to be precise) eventually petering out, dumping all that warm Mexican sunshine in the North Atlantic (Mexicans themselves are not allowed in without a visa). Other ocean currents do similar jobs, moving warm or cold water around the place and affecting local climates. The cold Benguela current originates down near Antarctica and runs up the arid west coast of southern Africa and past the Namib, one of the driest deserts in the world. While the cold water is partly to blame for the low rainfall in the region, it is also responsible for regular night-time mists that blow in and sustain what little life there is. El Niño – a circulation pattern of ocean currents and drought/flood events associated with it – gives us a taste of how small changes in currents affect the climate thousands of kilometres away. A temperature change of a few degrees in the waters off Peru not only plays havoc with the local fishing industry, but also causes floods and other "severe weather events" in the Americas, and crippling droughts in southern Africa and Australia.

The oceans are also the main source of water vapour in the air and our main source of rain on land, and how much water evaporates is a function of sea temperature. So it makes sense that warmer oceans will produce more water vapour and we will get more rain. It's not really that simple, especially when you want to know whether *you* will be the one getting wet. Like the oceans, the atmosphere is always moving – we call this wind. And because wind is driven by temperature and pressure differences, global warming may not bring the rain-clouds to the places it has in the past, or at the times or in the quantities we have become used to. All that the clever scientists will say is that it is highly likely that some areas will get wetter, particularly the high northern latitudes, and some may get drier, probably parts of Africa and Australia.

When warm, moist air rises, it cools and the moisture condenses into a cloud of tiny suspended water droplets called . . . a cloud. These small, scattered, ephemeral things that are the subject of so much art and poetry are a nightmare for our down-to-earth climate modellers. They disappear at the drop of a hat and pop up where they're least expected, happy to drift wherever the wind takes them. Scientists argue a lot about the effect that moisture and clouds have on climate and how to incorporate this into models. Clouds are good reflectors of sunlight, cooling the earth, but water vapour has a powerful greenhouse effect and contributes to maintaining heat in the atmosphere. Modellers have to simplify the effect of clouds on climate by "averaging" their impact, rather than trying to model the actual life of an individual cloud, otherwise the computers would go haywire. But the modellers have to be careful; such assumptions provide climate change-deniers with ammunition for trashing the model's predictions. To get around this, scientists use a variety of scenarios in their models on how clouds behave. Testing the model predictions against actual weather events, they are able to fine-tune the accuracy of the models.

There is a lot of energy contained in the water vapour that gets blown about, and when this energy is suddenly released blizzards, hurricanes and tornados develop. Remember your experiment with the kettle and how hard it was to heat air? Well, it's hard to heat air because almost all the thermal energy contained in air is, in fact, contained in its moisture content (in the water vapour). For example, air-conditioners work not only by cooling air, but also by drying it, thus removing its energy-rich moisture content. These evaporation and condensation processes involve the transfer of huge amounts of energy which also feed into producing "severe weather events". Global warming models predict that we will see fiercer tropical storms, hurricanes, and typhoons. Exactly when and where these will occur is a question of great interest to the insurance industry.

A final, and almost forgotten, role of the sea in climate is its ability to absorb huge amounts of atmospheric gas, and especially the one we are most interested in, carbon dioxide. During the nineteenth and twentieth centuries, the oceans absorbed almost half of the carbon emitted by humans. But the ability of the ocean to keep this carbon dioxide in a dissolved state is also a function of temperature. The warmer the water, the less dissolved carbon dioxide it can hold. And the less it can hold, the more remains in the atmosphere, and the greater its propensity to trap heat and warm oceans. Yet another pesky feedback effect.

Geographic factors

Local geographic factors are important for two reasons. Some of them play a role in determining global climate, while others, such as mountains, determine how global climate change will be felt in regional and local situations. Modellers must choose how much detail to incorporate in their models.

Large ecosystems such as the Sahara Desert or the Amazon Rainforest have created their own climates and even influence the global climate. Their vast size has an effect on the energy balance, i.e., how much of the sun's energy is absorbed and how much is reflected back into space. The Sahara gets so little rain partly because it is so dry and there is no moisture to form clouds.

The forests of the Amazon on the other hand give off so much moisture that it is thought that about 70 per cent of the rain that it receives is merely "recycled". The jungle is not just there because it rains every day; it also rains every day because of the jungle.

How do the models work?

By now you will probably appreciate how extraordinarily complex models are, given what needs to be incorporated in them (and we haven't even considered the human factor, like, for example, whether it would make any difference if Greenpeace ran the world instead of George Bush). So modellers need to simplify things, which they do by dividing the world into thousands of little boxes.

When you have a big, big problem to solve (and predicting climate change is an extraordinarily large one), the standard approach is to reduce the problem into small parts which can then be solved one by one. Climate models take this kind of approach. The other trick is to make as many "assumptions" (educated guesses, and these climate scientists are very smart after all) as you think you can get away with, or to ignore "variables" (things that keep changing) that you think don't have much of an effect. The more assumptions, and the fewer variables, the less complicated your model has to be, but, at the same time, the less accurate its predictions are likely to be.

(The next section is quite technical and you might want to skip some of it. If you do, we suggest you resume reading at the concluding paragraph.)

Currently, one of the most sophisticated types of models is known as a "3-Dimensional Atmosphere-Ocean General Circulation Model", or, simply, "GCM". There are different versions, but all of them start by dividing the oceans and atmosphere into thousands of imaginary boxes, stacked next to and on top of one another. The first layer of atmosphere boxes covers the surface of the earth. On top of this layer is another. And on top of that layer, yet another, and so on, and so on to the upper reaches of the atmosphere. Beneath the surfaces of the oceans there are also heaps of little, imaginary boxes. The smaller the boxes, the more of them there are, and the more complex the model becomes. The one used by the Goddard Institute of Space Studies in New York, for example, uses 3,312 boxes per layer. Typically the atmosphere boxes are between 250- and 400-km^2 (about the size of Lebanon or The Gambia) and about 1-km high. Underwater boxes are generally sliced a little thinner. These boxes are stacked twenty or more levels high, reaching into the upper atmosphere. Got the picture? Hold onto it. Things get worse from here.

So obviously each imaginary box (except for the bottom and top layers) is surrounded by six similar boxes, one on each of

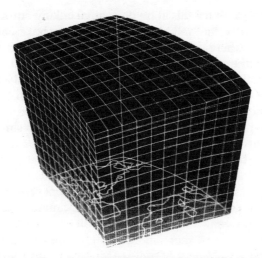

A representation of a 3-Dimensional Atmosphere–Ocean
General Circulation Model.

its four sides, and one each above and below. For each box, a
set of computer programs calculates how different weather
elements (temperature, air or water movement, water vapour,
dust, greenhouse gases, etc.) interact with one another over a
specified "time increment" (an hour, say), giving rise to a new
set of values for each of the elements in each box. These calculations use known laws of physics or experimentally observed
relationships. To make things more complicated, because the
lines separating the boxes are imaginary, what happens *inside*
one box is also partly influenced by the changes inside each of
its neighbours and vice versa. In addition to the imaginary
boxes in the air, the effects of ocean currents also have to be
modelled. In some parts of the world, oceans serve to cool
things down, while in others, they warm things up. Melting ice
and increased flows of freshwater into parts of the ocean
change temperature and salinity, all acting to strengthen or
weaken ocean currents. These factors have to be accounted for
in the individual boxes.

After the people in the white coats have switched on the

computer (don't try this at home – we're talking supercomputer here), specified the "initial conditions" for the boxes, made a few assumptions here and there, and specified the time increment (say, one hour), the computer does one round of calculations. While one round of calculations may simulate one hour of actual weather, the computer models it in less than a split-second. The output of the first round serves as input for the next round, and so on and on ad (almost) infinitum. When the supercomputer has simulated the changes in climatic conditions for 100 years in steps of one hour at a time in thousands of little interacting boxes, it can be gently brought to a halt, fed and watered, and asked for the answer.

While Ocean-Atmosphere GCMs are current state of the art, as computers get more and more powerful, models can become more complex and accurate, and still spit out the answers before Christmas. One of these more complex models is called the Integrated Assessment Model (IAM) which incorporates some GCM features, but also tries to include social variables such as demographics (how people settle or move around), land use, economics, etc. Because computer power is a significant limiting factor, they include these extras at the cost of losing some physical variables. But watch this space.

Since global warming models are built to predict global temperature in the future, how can you tell if they will make accurate predictions or not? What if you left out a critical X-factor, over-estimated the cooling effect of aerosols . . . or hit the minus key instead of plus? One way to *validate* the models is to use information from the past as initial conditions, and then run the model to see how well it predicts the present. Although getting climate information from the past is a bit tricky, modellers have managed to estimate from various sources like tree rings and ice-cores, and have found that . . . their models aren't bad! Historical predictions are not far off what actually happened. This makes it all the more important to take the models seriously with regard to what tomorrow may bring.

Models as decision-making tools
Models, for all their faults, can at least help us to understand the *most likely* consequences of our actions. Is global warming really going to be that bad that we have to switch off all the lights and

put the car up on bricks? Or can we wait until tomorrow . . . or next week? If we give up fossil fuel altogether will it make any difference or is it already too late?

One can debate all night whether human-generated carbon dioxide will cause a 1°C or 5°C increase in global temperature. In the morning we still have to try to do something about it. So, in theory at least, by designing the climate models to illustrate possible scenarios, economists, policy-makers and world leaders can chart a way forward that is in the best interests of humankind. In theory. How our leaders play their deadly game of double bluff with the future of our climate is discussed in chapter eight. So what is a scenario? In climate modelling terms it is a "structured account of a possible future". Modellers ask useful questions like, what if our populations continue to grow rapidly, as do our economies which remain dependent on fossil fuels? Or, what if population levels stabilize and we are able to de-link most of our economic growth from carbon? Or, what if we manage to do this but are unable to stop the destruction of tropical forests? Each scenario provides modellers with the basis on which to make assumptions about carbon dioxide emissions, the effects of carbon sinks and other elements critical to the models' predictions. They also allow us to understand why climate predictions are often so vague and contain such wide margins of error.

But despite their shortcomings, climate models are the best decision-making tools we have and we are lucky to have them. Archaeological evidence from around the world is beginning to show the huge and devastating impact that changes in climate have had on earlier civilizations. Consider the poor Akkadians. About 4,300 years ago they were living the good life in the fertile Euphrates valley, the breadbasket of the known world at that time. The mild and dependable climate meant that for a few hundred years they had managed to grow enough food to support a city which had grown to have a population of over 20,000 people. And then drought struck – one that lasted for the next three hundred years! Thousands were wiped out. The poor Akkadians had simply not seen it coming, and they'd only recently invented agriculture.

Thousands of years later, and on the other side of the world, the Mayan civilization came to an abrupt end. No one really knows why, but some archaeologists suspect a sudden and catastrophic drought, which the Mayans, with their slash-and-burn agriculture were simply not equipped to cope with. The Mayans had invented a five-thousand-year calendar and could predict astrological events with great accuracy. But about climate change, like the poor Akkadians three thousand years before them, they knew zip. At least we have our models to play with, and, if we're smarter that the Akkadians and the Mayans, we will use them to help us navigate the likely impacts on our way of life as we know it.

Chapter Four

᛫

Why Should We Worry About Global Warming?

This chapter examines what the models predict about future climate. These predictions are relatively straightforward on a global level, though considerably more complex on a regional level. What do the predictions mean for coastal communities, agriculture, diseases, bugs, and biodiversity? The model predictions are supported by observations of secondary effects such as the extinction of frog species, migrating quiver trees, Hurricane Katrina, and the melting of the permafrost. The chapter also examines the close historical relationship between human evolution and variations in climate.

Climate and evolution

Just how closely our climate is linked to the evolution of the human species and the development of "modern" civilization is the subject of many fascinating archaeological studies which we won't go into. Well, all right, we will, but just a little bit. (Creationists should skip this section, unless looking to change their minds.)

Fossil records show that way back when, our ancestor ape-people in Africa had gone about their ape-people business relatively unchanged for at least a million years. Half in the trees and half on the ground, they were neither fully ape nor fully human, happy to swing through the trees and subsist on a diet of fruits and nuts. About three million years ago, slowly but surely, the climate began to change. Huge ice sheets started growing in the northern hemisphere, creeping slowly southwards. The temperature of the North Atlantic dropped by as much as 25°C, causing cooler, drier air to blow over the African continent. African forests shrank and gave way to open savannah. Of course these changes happened much too slowly for any individual ape-person to recognize, and, anyway, life was tough enough and their brains were only the size of a tennis ball. Then in the middle of this (relatively) rapid change in the climate came a (relatively) rapid and remarkable change in the ape-humanoid. Fossil remains from less than two million years ago show the "sudden" emergence of ape-people with significantly larger brains. Walking upright had become the preferred method of getting about, teeth seemed less suited to a solid diet of forest fruits and nuts, and there is evidence of sophisticated stone tools having been used. Survival in the new climate and the dramatic changes it wrought on the landscape, so the theory goes, required a large brain and a rapid stride. Thanks to climate change we learned to think on our feet, and that's what we've been doing ever since – not that it seems to have made much difference!

And much, much later (another theory goes), thousands of years of unbroken, warm, moderate climate allowed those big-brained walkers to populate almost every corner of the planet. We bred like humans and eventually had to settle down, plant crops, and raise animals. Another competing theory proposes that it was the vagaries of our current climatic regime that, in fact, forced us to invent agriculture to provide a stable food source. What really happened doesn't matter. The point is that past climate has been a key evolutionary force in shaping human civilization. And because of its stability we have been able to bring civilization to its peak – the combination of convenience store, TV remote, and couch. Are you ready for the next great climate–driven evolutionary leap? Or will you just change the channel?

What do the models predict?

In building and running climate prediction models, the modellers include elements about which they are very certain (for example, warmer oceans evaporate faster and at a known rate), and elements for which they are rather less certain (e.g., how strong various feedback effects will be), and variables which underwrite the different scenarios that they describe (e.g., how much carbon dioxide we will emit). The final predictions that the models make, therefore, contain some in-built uncertainty. This is not to say that there is any uncertainty about the effects that a model predicts, but rather about their severity, precise timing, or such-like.

Climate models make a lot of sense to climatologists, but how are they used to give policy- and decision-makers something to think about? The IPCC has developed a range of scenarios which illustrate how the world may respond to future development challenges. Each scenario gives an indication of what assumptions to make and what to feed into the climate models. There are some obvious key variables which modellers think will make critical differences to global warming outcomes, such as population growth, economic growth and how wealth is distributed, the degree of technological advancement, international co-operation, social and environmental equity, and of course, the degree to which we remain dependent on fossil fuels. Most of these factors are interrelated and influence each other, so to build them into a sensible model, the IPCC has generated what they call "storylines" which help them to group certain assumptions. Although the IPCC has developed six different storylines which include about forty different scenarios, the Fourth Assessment Summary Report highlights only six of these as being particularly illustrative for policy-makers. They are described here in brief to give you an idea of how the science is translated into a few "what if" situations that we can make sense of.

The A1 storyline describes a future world of rapid economic growth, a global population that peaks in the mid-century then declines, and a rapid advance in new efficient technologies. There is a political and economic convergence across the world, with increased cultural and social interaction and increased economic equity. In other words, we all get along well and sort out our problems as a global community. Within this storyline, scenario A1FI assumes we are still *highly dependent on fossil fuels*, scenario A1T assumes we have *moved substantially to non-fossil fuel energy* sources, and A1B assumes something of an *in-between* state.

The A2 storyline and scenario family describe a very *heterogeneous* world. Each country/region looks out for itself. Self-reliance and preservation of local and national cultural identity are the underlying themes. Population growth in parts of the world continues. Economic development is confined to certain regions and technological development and transfer between regions is slow and fragmented. Co-operation on global problems could be better. Sounds a bit like the world we know, doesn't it?

In the B1 storyline and scenario family, population peaks in the mid-century and declines. Globally there is *strong social and political convergence*, as in the A1 storyline. We all get along and help each other out. There is a rapid change worldwide towards a service and information global economy which is less material- and energy-intensive. *Clean and efficient technologies are widely used.* There is co-operation on global solutions, and an emphasis on social and environmental sustainability and improved equity, but not necessarily because of global warming.

Finally, the B2 storyline and scenario family describe a world in which the emphasis on environmental and social sustainability is strong, *but localized rather than global*. It's a sort of socially and environmentally friendly version of storyline A2. The population in some regions continues to grow, while economic growth and technological advancement is slower.

The direct global warming impact we can expect to see by the end of the century based on these scenarios is given in the table below. Which scenario do you think will play out? Can you spot the one that will cook us fastest?

IPCC scenario predictions for 2100	Temperature change ^oC relative to 1980/99 average		Sea level rise cm relative to 1980/00 average
	Best estimate	Likely range	Likely range
A1F1	4.0	2.4 – 6.4	26 – 59
A1T	2.4	1.4 – 3.8	20 – 45
A1B	2.8	1.7 – 4.4	21 – 48
A2	3.4	2.0 – 5.4	23 – 51
B1	1.8	1.1 – 2.9	18 – 38
B2	2.4	1.4 – 3.8	20 – 43

In climate change, the IPCC's best-case scenario (B1) predicts global warming *best estimate* of 1.8°C by 2100, but *likely* to be in the range 1.1°C and 2.9°C. The worst-case scenario (A1F1) predicts a *best estimate* of 4.0°C increase but *likely* to be in the range 2.4°C to 6.4°C. Notice the term "likely" and the range of the prediction. This may look like they are toying with us, hedging their bets, or are just plain incompetent. But, no, this is an expression of the statistical probability of a specific outcome actually happening. "Likely" means there are at least two chances in three of it happening.

The consequences of smoking are similar. As a smoker you know that you have a shorter life expectancy than your non-smoking friend. But you don't know exactly when or how painfully you will die. You might even live longer than your friend. Your friend might die of lung cancer. But *on average*, smokers die younger than non-smokers. They are also more likely to die of smoking-related illnesses.

The reason for belabouring the above point is because all along you've been burning to ask The Big Question: What difference will global warming make to my life? Well, we're sorry, but despite costing millions to develop and running for weeks and weeks on massive supercomputers, there is a real limit to the specific predictions that models can provide. They can tell us with a statistically high degree of accuracy, for example, what the average global rainfall may be, but they may not be able to tell you with much accuracy where, or when (or how hard) it will fall – which is the kind of information we're all really hoping for.

Nevertheless, the following section examines the effects that climate change may have in some key areas. Some of these effects may not occur; some may be milder than described, while others may be worse. The truth is that we don't yet really know.

Food security

The plants that we depend on for food generally require a warm climate, carbon dioxide for photosynthesis, and sufficient water. All the climate models predict these requirements in abundance, so . . . it looks like we're in for a bumper crop! Not really.

Some areas may indeed benefit, particularly at higher latitudes where agriculture is currently constrained by a short growing season and cool weather. Some crops may be able to grow in previously unimaginable locations ("Pass me those delicious Norwegian olives, darling."). The problem is that there are additional factors influencing food production. For example, ozone, which is increasing due to pollution, particularly in the northern hemisphere, reduces crop yields significantly.

Regions that are already warm, will get even hotter. Rice is particularly sensitive to temperature, and yields are predicted to drop. More carbon dioxide in the air may benefit some crops but not all, and recent studies indicate that its effect has been over-estimated. The increased rainfall that models predict may not fall in the right places, at the right times, or in the right quantities. Also, the predicted increase in rainfall will be as a result of increased evaporation. Over the sea, this is fine. On land, in areas which are already dry, evaporation is a bad thing because it sucks precious moisture from the ground. It is very difficult to predict how increased rainfall and temperature will be distributed regionally. Some models indicate that Africa will experience longer, dryer, dry seasons and shorter, wetter, wet seasons. Changing climate will also mean that some agricultural pests, moulds, and other plant and animal diseases will move into new regions, or will become more of a problem.

Predicting how fast climate changes will occur with any

certainty is difficult. An increase of 4°C will cause global yields of cereal crops to drop by 11–20 per cent. If this change is slow and steady, farmers and agricultural researchers may have enough time to develop new, more resilient varieties. Whatever the pace of change, Northern farmers with better infrastructure and greater economic resources at their disposal are more likely to be able to respond to climate changes than their poorer Southern counterparts. Countries that are currently "food stressed" are likely to become even more stressed. So even if overall global productivity is maintained, this will be of little comfort to the Kenyan farmer in 2100 when his small plot of maize is almost drowned in the wet season, decimated by pests a few months later, and finally wilts in terrible heat.

Small island states lose their "state-us"
That sea level will rise is not disputed. It has already risen (by about 17 cm over the last 100 years) and will continue to rise. The big question is how fast it will rise and by how much. The IPCC's scenarios estimate a rise by the end of the twenty-first century of between 18 cm and 38 cm in the best case, to between 26 cm and 59 cm in the worst case.

This doesn't sound like an awful lot; nevertheless, high tides will become just that little bit higher, and storm surges will surge that little bit more. When the sea level rises it doesn't just go up, it also comes in, depending on how flat and low-lying the land is. This can't be a good thing unless one lives in a boat.

With coastal flooding already affecting some 48 million people each year, things can only get worse. Some 40 per cent of the world's population lives within 100 km of the coast, with 100 million people living less than a metre above sea level. The IPCC estimates that a 40 cm rise (admittedly at the higher end of the predictions) will put 75 to 200 million people at risk of annual flooding – unless they get out of the way in time. Vast tracts of highly productive, densely populated, low-lying river deltas may disappear altogether. About 17 per cent of Bangladesh, for example, may be flooded. For richer Northern countries, getting out of the way and building protective dykes and barriers will be disruptive and costly, but feasible. Moving a few million Bangladeshi people off the fertile Ganges delta will probably lead to that country's social and economic collapse. And the effect won't be localized as millions of environmental refugees seek new places to live and challenge others to respond to their humanitarian crisis.

The small island states will be particularly hard hit – some may disappear altogether. Their challenge is twofold, particularly for those on coral atolls, the highest point of which may be less than a metre above sea level. The problem is not only that waves may wash over them, but that the fossil coral beds that they rest on are very porous. Building sea defences will be costly, and ultimately useless. If the sea doesn't wash over them, it will rise up and swamp them. The highest point in the group of nine Pacific coral islands that comprise the nation of Tuvalu is only three metres. Island leaders have already conceded defeat in the face of rising sea levels and have made plans to begin evacuating the approximately 11,000 inhabitants to New Zealand (having been rebuffed by Australia) and other, higher neighbouring islands. Tuvalu is paying the ultimate price for the rich world's experiment with global warming.

Salt water intrusion will also result from the sea level rising and coming in. Rivers and coastal aquifers would become salty, threatening drinking-water supplies in many coastal towns and cities. It is not only people that will be affected, but investments in coastal infrastructure may be lost or may have to be protected at great cost. The impact of sea level, which is devastating and costly for all countries, will be softened for some and magnified for others due to their capacity to plan and pay for it. As with food security, it is the richer nations that will be able to adapt better.

How fast will this happen? Sea level rise has two causes: the expansion of oceans as the water warms up, and the release of glacial ice. Unless there is runaway melting of the Greenland ice cap, which is not (yet) likely, we will continue to see a slow and steady rise. Remember too, that because of the inherent thermal inertia of the oceans, rising sea levels will be with us for centuries after we have managed (we hope) to stabilize and then reduce greenhouse gas levels. If the unthinkable happens and Greenland melts, our coastlines will be completely unrecognizable following a 7-m rise in sea level.

Human health: bad air and muddy waters

Bad air (or *mal aria*) used to hover above the English fens and other swampy places causing raging fevers and, sometimes, even death. The fens were drained, and the mosquitoes carrying the fever either moved elsewhere or died. Fewer than ten years ago, *Anopheles funestus*, a malaria-carrying mosquito resistant to the pyrethroids used to control the more common *Anopheles arabiensis* made a comeback in South Africa. Mosquitoes only bite in the evening and prefer to eat indoors so residual spraying of DDT on the inside walls of homes and animal sheds is used to control it (with approval from the World Health Organisation). Both the warmer weather and wetter conditions predicted by global warming will help malaria-carrying mosquitoes to breed more vigorously and may result in their spread into temperate regions and higher altitudes, which will no longer be the refuges from the disease that they once were. This will have a far worse impact on people in countries with a limited capacity to control malaria. It will also increase the need to use pesticides.

Rainfall, temperature, and humidity have a major influence on the distribution of other pests, parasites, and pathogens as well. Lyme disease, bilharzias, Rocky Mountain spotted fever, and tick-borne encephalitis may also spread. It is not known what the effect of a warmer world will be on the spread of "new" diseases such as SARS and bird flu, but it is likely that, once again, poor people in countries lacking the infrastructure and funds to treat and control the spread of diseases are likely to be the worst affected.

The impact of changing rainfall patterns and extreme weather events such as droughts and floods will continue to have a huge impact on peasant agriculture, threatening household food security. When people don't have enough food, their resistance to disease is much lower. Water-borne diseases such as typhoid, cholera, and dysentery thrive during floods; and droughts bring diseases associated with poor water quality and inadequate sanitation. Those already barely surviving will be very badly hit. Flooding exacerbates these problems because it becomes much more difficult for people to get to clinics, or for health workers to get to distant villages. Immunization campaigns and other public health measures could be disrupted.

A final health impact of global warming is the predicted increase of respiratory diseases due to air pollution arising from ozone and volatile organic compounds formed in urban areas and large forest fires outside of cities.

The impact of diseases on public health depends on the general standard of living, the level of access to medical infrastructure, and government's ability to control the spread of diseases.

"Thar she blows!" – Extreme weather events

Hurricanes and tornadoes are best viewed on TV when they are happening on the other side of the world. Deep down, everybody is fascinated by natural disasters, but climatologists, being objective, prefer to call them "extreme weather events". Global warming is predicted to bring more of these and and of greater intensity as a result of the increased thermal energy that an increase of a few degrees of temperature represents, and because higher temperatures will cause increased evaporation and more water vapour in the air. In 2004, the South Atlantic had its first observed hurricane, which hit Brazil.

While our climate models get more and more sophisticated, they are, unfortunately, still not able to predict actual weather events. So where, when, and how extreme future events will be, it is impossible to say – at least until a few days before they hit. But if the proper planning and emergency response strategies are in place, with a few days warning a country can at least minimize the loss of life, if not the damage to property. Again, this is easier in countries with the appropriate infrastructure and in which people are not living in areas already exposed to floods and droughts. Hurricane Katrina, however, showed that even a very wealthy country finds it difficult to cope with extreme weather events. The combination of an extremely large hurricane with a woefully inadequate response strategy had a devastating impact on the lives of the people of New Orleans.

Fires are also predicted to increase as a result of global warming due to shifting rainfall patterns and increased evaporation from soil. When forests burn, the smoke can travel for thousands of miles and affect millions of people who breathe in the particulate matter. Forest fires also burn carbon and release millions of tons of carbon dioxide into the air, thereby increasing global warming and the probability of more fires.

Biodiversity
Plants and animals have evolved over millions of years to suit particular habitats, where conditions are just right for them. Some species are generalists, able to adapt to a range of conditions and habitats – cockroaches and crows come to mind. At the other extreme, some species have evolved to occupy such specialized habitats that they are found only in a single forest, valley, or mountaintop. Gradual changes in the climate will cause those species that can to "migrate", or, if they can't do that, simply to disappear. Those species that cannot migrate will experience not only deteriorating conditions, but also increasing competition from new immigrants. Plant species, particularly, do not migrate easily. Not only are they not as mobile as insects and animals, but human developments like agriculture, towns, and cities will act as barriers. The fate of animals and insects that have evolved to depend on a handful of plant species will be tied to the latter's ability to migrate to more suitable climes. Migrating upwards presents another challenge. Lowland ecosystems, connected to each other, will become vulnerably isolated islands as they move up mountain slopes.

Climate models predict that climatic zones in the mid-latitudes could shift towards the poles by 150 to 550 km during the next 100 years. Warmer temperatures also mean that higher, cooler altitudes will become warmer. A zone shift of between 150 and 500 m in the next 100 years is predicted. Some of these shifts have already been observed in the behaviour of plants and animals. An analysis of studies of more than 1,700 species has indicated significant range shifts of over 6 km per decade towards the poles, and of over 50 m per decade in altitude. Natural springtime events are occurring roughly 2.3 days earlier each decade.

Ocean ecosystems will also be affected given that huge amounts of global warming is absorbed by the sea. The coral reefs of the tropics are not just playgrounds for the rich and flippered, but are also sites of exquisite marine biodiversity and nurseries for countless species of deeper water fish on which we depend for food. Coral reefs are a symbiotic home for coral polyps who build the tiny skeletons, and specialized algae which provide food through photosynthesis. Increasing ocean temperatures will turn this mutually beneficial relationship ugly, as the algae leave for cooler living quarters. The polyps, unfortunately, are not as mobile. Without their algal partners they lose their colour along with their source of food and they bleach, and die. A 2°C increase will bleach 97 per cent of the world's coral reefs resulting in a terrible toll on the fish and other marine animals that depend on them.

And, on the other side of the world, in the frozen north, the available habitat for the remaining few thousand polar bears is steadily decreasing as the Arctic sea ice melts and retreats. The polar bears have nowhere else to go. Some scientists estimate that global warming may cause the extinction of between 15 and 37 per cent of all plant and animal species by 2050. Some of our fellow creatures, like the golden toad of Costa Rica, we will barely have the opportunity to get to know before they are gone. Others are well known for the role they play in providing a genetic store for agricultural plants and other medicinally "useful" plants and animals. But trying to put a value on such a massive loss of species and such a devastating rupture of our planet's web of life seems ridiculous. The earth's biodiversity is irreplaceable, and the wider impacts of such devastation are hard to imagine. If ever there was an argument for applying the precautionary principle to global warming, this is it.

Section Two

The Politics: How the World is Responding

Chapter Five

Adapt or Simmer

Although there are already small-scale, localized responses to climate change, these can only go so far without political will and co-operation on both national and international scales. We need to find ways to adapt to climate change because we have already set in motion changes which will continue for the next fifty years or longer; but we also need to find ways to mitigate climate change.

Those who think deeply about global warming have classified our response to the crisis into two categories: mitigation (how are we going to stop pumping greenhouse gases into the atmosphere?); and adaptation (how are we going to survive this disaster?) This section will look at adaptation, while we will examine mitigation in the following chapter.

We are all constantly adapting to many things – to some of them consciously, to others, not. Adapting to change is a fundamental trait of all living things, and humans and other species have certainly adapted well to slowly changing climates over the millennia. But like any other living species, there are physical limits to our adaptability. There is also a limit to how quickly we can adapt, and there are other limits, too. In our hunter-gatherer past, it was possible to adapt to changes in the climate simply by moving on to where the grass was greener. But that option is no longer available to us as the land and resources are simply not there any more. Besides, it is difficult to imagine moving a world of 6 billion or more people, half of whom are rooted in, and dependent on the economies of towns and cities.

It is the wealthier members of society who will be able to adapt better, and those who are already poor and vulnerable who will bear the brunt of the impact of global warming. Unless we want to risk deepening these divisions in society, we need to think about adaptation as preparation. So the challenge is how to adapt in a conscious and planned way, and we have to begin now.

Adaptation is a catch-up game. Even if all greenhouse gas emissions were stopped tomorrow (in our dreams!) we would still be sitting with an atmospheric carbon dioxide concentration well above pre-industrial levels. And because of the lag-effect of the oceans and the slow natural uptake rate of carbon dioxide, global warming would continue for many years to come. What we adapt to today is the global warming effect of greenhouse gases released decades ago.

One of the clearly predictable changes will be a rise in sea level and more extreme flooding. The Dutch are past masters at the art of taming water. About a quarter of their country lies below sea level, while another quarter is so low-lying that, without protection, it would be regularly flooded. The legendary sea and river dykes of the Netherlands (all 15,700km of them) and the windmills (now replaced by fossil fuel-powered electric pumping stations!) have held back floods and high tides for hundreds of years. A natural response to a rising sea level would be to build more and bigger dykes, install more and bigger pumps and reclaim more land from the sea. But some in government insist that new thinking is required, because, when the dykes fail, as they inevitably must, the consequences would be disastrous.

The new thinking is that living with the elements would,

perhaps, be more sustainable than fighting them. One innovation is an experimental "amphibious house" which rests on hollow concrete pontoons that are heavy enough to provide a firm foundation on dry land, but light enough to rise above the floodwaters. More radical solutions suggest that rivers and canals should be widened to cope with larger flows, and certain rural areas should be set aside to flood naturally, acting as buffers to protect more populated urban centres. Naturally these ideas go against the grain for most Dutch people, and putting them into practice would be a political hot potato.

On the other side of the world in South Africa, in the arid, rocky north-west of the country, descendants of people who were once slaves of Dutch settlers in the Cape grow small plots of indigenous rooibos tea. For the past few years they have been meeting every three or four months with climate change academics. The farmers share with each other and with the scientists, what their weather predictions are for the coming season and what sort of adaptation measures they might take – like increasing mulch, planting earlier or later, or stockpiling tea for the future. The scientists in turn, share with the farmers their knowledge of climate change and their long and short-term predictions.

The same farmers participate in a project to develop and manage a "biodiversity corridor" stretching from the coast, over the Cedarberg mountains and into the arid Karoo. If the plant and animal ecosystems of the Cape fynbos (one of the world's six plant "kingdoms") is going to survive, species must be able to migrate. Isolated nature parks and reserves will no longer do the trick. Farmers will have to facilitate linking them up.

The global insurance industry is also adapting, although perhaps somewhat ungenerously by raising premiums on weather-related insurance and lowering their liability limits. The philosophy of insurance is to protect against the improbable. More of what used to be improbable is now becoming certain. Historical data is becoming less reliable as a tool for forecasting future risk. It is not only because insurers think that natural disasters due to global warming will increase (can we still even call them "natural" disasters?), but also because we are becoming more vulnerable to them. As populations increase and people settle in areas previously regarded as unsuitable, and as our societies become more and more dependent on centralized economies and services, so the cost of recovering from "great weather-related catastrophes" (as they are known in the language of insurance companies) increases, and somebody has to pay.

Insurance companies also see good business opportunities in global warming. Industries which find it technically or economically unfeasible to limit their own carbon emissions may invest in projects which reduce carbon (planting forests, installing solar heating in houses, etc.) and thus earn themselves "carbon credits" (there is more about carbon credits in chapter nine). But if those industries' forests burn down, or their solar heating projects do not work, they face a financial penalty. Now it is possible to buy insurance against one's investment not delivering the carbon credits that it should!

Local authorities in some parts of the world are adapting by revising building codes, redrawing 100-year flood lines, and re-thinking water supply infrastructure.

But there are things we should be adapting to that we are not. Why? One reason is that we simply can't feel the heat, yet. We keep our buildings cool using thick concrete walls. We insist on travelling long distances by aeroplane – to conferences which protest against government inaction, or to snorkel among the last remaining patches of coral reef. When, because of climate change, there is not enough snow to ski on, we *make* snow, by switching on fossil fuel-powered snow machines. It's madness!

It is no secret that those most vulnerable to the negative effects of global warming are also those least able to afford it, and who have contributed least to the problem. It seems only fair, then, that those who have caused most warming should also pay for poor countries to adapt. They do, but, as ever, with many strings attached. Under the Framework Convention on Climate Change it was agreed to set up funding mechanisms for adaptation. One such mechanism is supposed to be replenished mainly through a 2 per cent levy on Clean Development Mechanism Projects (there is more on these in chapter nine), but is not expected to generate significant resources until after 2010. Three other dedicated funding mechanisms, the Strategic Priority on

Adaptation Fund, the Least Developed Countries Fund, and the Special Climate Change Fund are supported by a range of donors. The latter two funds attracted only $43 million in 2005–6 whereas the World Bank estimates the cost of adaptation measures could run into tens of billions of dollars. By contrast, France spent $748 million upgrading its hospital emergency services after the 2003 heatwave. Some have estimated that the global fossil fuel industry is subsidized to the tune of $235 billion per year.

In short, the money is clearly inadequate. What about the strings, then? The distribution of funds is currently managed by the Global Environmental Facility (GEF), a body which also channels donor funds to biodiversity and desertification projects. Needless to say, decision-making in the GEF is dominated by Northern interests – he who pays the piper calls the tune. Worse still, the World Bank is a joint fund manager – another sore point with many developing countries who fear, with good reason, that the Bank will use its position to push its own political agendas.

The only real, effective, long-term adaptation, however, is mitigation. If human beings are to survive – and it really can be put as plainly as that – we simply have to reduce the amount of carbon dioxide and other greenhouse gases which we pump into the atmosphere.

Chapter Six

Reducing Emissions
(and perhaps capturing some)

This chapter will explore what constitutes a dangerous level of anthropogenic interference and, therefore, what level of mitigation is needed. It looks at some options and measures for reducing emissions and discusses capture and storage options, as well as possible ways of reducing warming – mirrors in space, for example! We cannot count on future innovation, but should be open to it.

". . . asking whether it's practical or not is really not going to help very much. Whether it's practical depends on how much we give a damn."

Robert Socolow, asked by Elizabeth Kolbert whether he thought stabilizing emissions was a politically feasible goal, in an article, "The Climate of Man – III: What can be done?" in the *The New Yorker*, 9 May 2005

All but the most hardened spindoctors believe that something must be done to stop carbon dioxide concentrations from reaching "dangerous levels" (and perhaps even the spindoctors believe it, but they need the work). But what is a "dangerous level"? How close to this are we already? How much time do we have to act? And what is to be done?

Remember the Keeling Curve in chapter two? This was the first clear indication that carbon dioxide levels in the air were steadily rising. Lots of things have happened since then, but concentrations just keep going up. We know that carbon dioxide levels have already caused some impact on temperatures and the sea level. But at what point should we begin to panic? When does carbon dioxide become "dangerously high"? Scientists, politicians,and activists have bandied various numbers about – 400 ppm, 450 ppm, double pre-industrial levels, and so on. It may seem that there is no consensus, but that is not the case. The problem relates more to the difficulty of pinpointing such a number with a high degree of accuracy given what we currently know; and the absurdity of the question itself. Dangerous in what way? And for whom? Melting polar ice is dangerous for polar bears and their Inuit hunters, and it is happening right now. If we carry on with carbon dioxide levels increasing at the current rate for the next fifty years and then stabilize at a steady 51 gigatonnes of carbon dioxide per year for the next fifty years (not an unlikely scenario), we will end up with a greenhouse gas concentration of more than *three times* that of pre-industrial levels. There is complete consensus that that would be *very dangerous indeed*.

Looking a little more closely at what lies behind many of the numbers, it seems that they are a fudge between what is likely to be a point-of-no-return level according to climate change models and what is politically acceptable. In 2005 Sir David King, the British government's chief scientist, proposed a target of 550 ppm. When challenged that at this level there would only be a 10 to 20 per cent chance of keeping global warming under 2°C, he replied that a lower target would lose him credibility with the government. Kolbert argues that the more we learn about global warming, the lower the number becomes. Most scientists believe that exceeding 450 ppm would be dangerous. Others bring the number down to 400 ppm – not far off our current concentrations. There is also a fudge between carbon dioxide concentrations and greenhouse gas concentrations *equivalent* to carbon dioxide – while carbon dioxide levels stand at 379 ppm, including all the other greenhouse gases brings that up to an *equivalent* of around 440–450 ppm, the level which most scientists think is dangerous. A sobering fact.

The IPCC Fourth Assessment Report notes that to stabilize the air's carbon dioxide level at 450 ppm, we would need to reduce our cumulative emissions of greenhouse gases over the next 100 years to 1,800 gigatonnes. If we continue to pump out what we did last year, every year for the next 100 years we would emit 2,460 gigatonnes. And remember that despite efforts to curb our behaviour, each year we pump out *more* than the year before.

Further research is likely to bring greater certainty as to what a "dangerous level" of greenhouse gases is, and as this comes to light it can and must inform strategies to reduce emissions. But playing around with projected numbers only delays the inevitable, and the more we delay, the more difficult and expensive the problem becomes to solve. We already know enough, and we know that we need to do something now, without delay.

Keeping the concentration of greenhouse gases within a particular range means we have to limit how much we emit, and this must be less than or equal to the amount that the earth's natural processes can absorb. As the natural sinks fill up we will have to emit less and less. The sea absorbs a significant quantity of carbon dioxide, but the oceans will be able to absorb ten per cent less carbon dioxide in 2100 than they are able to today as they "fill up" with dissolved carbon dioxide. Even maintaining current emission levels requires a great deal of effort. Reducing them is an unimaginable challenge.

Because the task is so daunting, a number of people have tried to break it up into more digestible chunks. In a journal paper in *Science*, Pacala and Socolow identify "stabilization wedges", interventions that would each prevent 1 billion tonnes of carbon

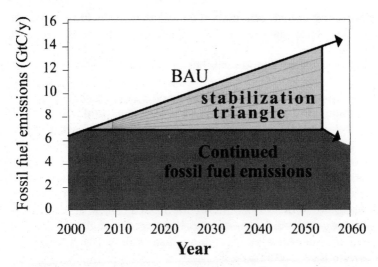

Pacala and Socolow's stabilization wedges. The top line shows projected
"business as usual" (BAU) emissions, while the bottom line shows reduced
emissions as a result of the impact of various stabilization edges.

(the equivalent of 3.67 billion tonnes of carbon dioxide) from
being released into the atmosphere over the next fifty years.
These are interventions for which we already have the
technology. They are called *stabilization* wedges because if we
put the right number in place they will stabilize our emissions.
The diagram shows why they are called *wedges*. To maintain
current levels of approximately 7 GtC per year, Pacala and
Socolow suggest we need seven wedges. To *reduce* emissions we
would obviously need more. George Monbiot maintains that
Britain needs to cut 90 per cent of its emissions by 2030 and in
his book *Heat* illustrates how this could be done in some key
sectors.

In essence, the things we can do to reduce carbon dioxide
concentrations fall into four categories. Most of them can be
done at both individual and national levels: using less energy
(e.g., deciding not to fly); improving energy efficiency (e.g.,
driving a more fuel-efficient car); using energy that has a lower
carbon content (e.g., converting from coal to solar power);

capturing and storing carbon dioxide emissions (e.g., in permanent new forests)

Let's get practical and see what can be done.

Avoiding emissions by using less energy
(aka demand-side management, or conservation)

Imagine it is the year 2054 and the world's population has stabilized at around nine billion people who drive approximately two billion cars. If everybody drove his or her car an average of 8,000 kilometres each year instead of 16,000 kilometres (today's average), we would save 3.67 gigatonnes of carbon dioxide from being emitted. This is one of the Pacala–Socolow wedges. Of course we would also save another wedge if there were one billion cars instead of two billion. Just think how much we could save if there were no cars.

Sensible urban planning and public transport is one of the best ways to reduce our reliance on cars. Only very few people would choose to be stuck in a traffic jam if they could instead get to work cheaply, safely, quickly, and pleasantly on a train, bus, bicycle, or on foot. Local holidays and excursions could also be made on public transport instead of by car, although the smaller numbers of people involved and our love of "the open road" would make this harder. Still, with innovative planning even that might be possible. But let us at least begin to get our cities to provide decent public transport and thereby eliminate the one-person-one-car daily commute.

Long-distance travel is particularly problematic. Aeroplanes, high speed trains, and ocean liners are major contributors to global warming. If everyone in the world were to be limited to an equal share of emissions below the level that would result in a dangerous concentration of carbon dioxide in the atmosphere, an individual would use up his or her entire quota for the year in flying from London to New York and back. This is because of the distance of the journey and the speed of travel. Worse still, because planes spew out other gases along with carbon dioxide, to calculate the full impact, one would have to multiply the carbon dioxide by 2.7. As Monbiot puts it, "If you fly, you destroy other people's lives." Think about that the next time you plan a trip.

There are already a number of other ways in which we can easily use less energy. When making a cup of tea, we could boil only the amount of water that we need, instead of filling the kettle. We could turn down the thermostat on our central heating slightly and wear a jersey instead. We could switch appliances off at the wall, rather than leaving them on standby – and write to the manufacturers to complain about this wasteful feature. Hotels frequently leave TVs on standby so that poor, exhausted guests have only to grab the remote control before flopping onto the bed. Is it really asking too much of guests to switch on the TV itself first? There are many other simple, effective steps we could take, simply by modifying our behaviour.

There is much that industry, too, could do. In many countries electricity produced by burning fossil fuels is ridiculously cheap for big business. It continues to be subsidized to create incentives for investment and as a stimulus for economic growth. Similar levels of incentives for industries that use less energy are seldom seen, though they should be. It is a difficult economic transition for a country to make. Some economies have successfully shifted from a manufacturing to a less carbon-intensive, services-based base, but only because their energy-intensive goods are produced in developing countries. Industrialization has been the means of wealth creation for developed countries, and it is unclear what kind of development is possible without this stage. Developing countries are therefore reluctant to give up energy-intensive manufacturing without a viable alternative way in which to generate wealth. Subsidized energy for big business needs to be phased out, but this will have to be done with careful consideration for developing countries particularly.

Improving design efficiency

Closely related to energy conservation is improving energy *efficiency* through better design and practice, with the net result that less energy (and fewer carbon dioxide emissions) produce the same result. The technology is well understood, and, in most cases, not very expensive. It is easy to design and build a house or office block to be warmer in winter and cooler in summer. In many parts of the world, traditional designs do this anyway. Thatch-roofed mud huts in rural South Africa, for example, are oriented and painted in a way that reflects the hot summer sun and absorbs the low winter sun. Needless to say, "progress" has seen this traditional technology being replaced by "modern" tin-roofed buildings that are extremely energy inefficient and costly for families to heat and cool. But energy-efficient design is making a comeback and there are schools of modern architects, engineers and builders working on "green" buildings. The Rocky Mountain Institute website (www.rmi.org), among others, offers practical tips on ways to "retrofit" your own home; it is much cheaper and easier, however, to build with energy-efficiency in mind in the first place. Revising existing building regulations to encourage passive heating and cooling would be a great help. For example, in sunny climates it should be mandatory for all new houses to be fitted with solar water heaters. Some countries, like Israel, require solar water heaters in all new buildings; this has resulted in the installation of around 50,000 new heaters each year. Roughly 70 per cent of all buildings in Israel now have solar water heaters.

One of the Pacala–Socolow wedges calls for a doubling of the fuel efficiency of motor-cars. In fact, we can already choose between buying a small, fuel-efficient car or a gas-guzzling 4×4 Hummer. Sadly, more and more people who can afford them are opting for a 4×4 as the only way to get their kids to school. Better fuel efficiency seems like such an obviously good idea, but it has a flip-side. If we can drive twice as far on one tank in a super-efficient car, will we use less fuel? Or will we go on more and longer trips because the cost is the same? Unless fuel costs also rise, or some other incentives are in place, we could end up emitting even more carbon dioxide and other noxious gases. Compare this with a friend (it's always someone else, isn't it?) who can justify eating a bar of chocolate on the grounds that she has just eaten a low-fat yoghurt. Sound familiar? To avoid such perverse logic, supplementary regulation is required to ensure that efficiency gains actually help to reduce climate change.

What is true for each of us and our cars is also true for industry. There are many good housekeeping practices and new technologies that allow manufacturing industries to use energy more efficiently. This saves energy but also reduces the cost of production, possibly stimulating more production, and consequently more energy use and greenhouse gas emissions.

Let's take a look at the source of the problem. Converting fossil fuels into electricity is a hugely inefficient process. Coal-fired power stations can waste up to two-thirds of the energy contained in the coal they burn. The electricity is transmitted on power-lines, sometimes over long distances before it reaches a home or a factory, incurring further losses. In the words of Amory Lovins: "The laws of physics require, broadly speaking, that a power station change three units of fuel into two units of almost useless waste heat plus one unit of electricity . . . At least half the energy growth never reaches the consumer because it is lost in elaborate conversions in an increasingly inefficient fuel chain dominated by electrical generation." Newer power station designs and transmission technologies are more efficient, but given a power station's lifespan of thirty to forty years, introducing these new designs is a slow process. At the end point, electricity is converted back into heat, or motion. Heating a pot of soup on an electric stovetop wastes even more energy, especially if the stove is an old one and the pot is the wrong size, or has an uneven bottom. Newer technologies at both ends of the transmission line tend to be more efficient in the way that they convert heat to electricity, or vice versa. New technologies also provide opportunities for using waste heat in more productive ways – by piping hot water to provide heating to factories or houses, for example. Similarly, large industries that use coal or oil-fired boilers to produce steam for industrial heat, may be able to generate electricity "on the side" when they have excess steam, and feed it back into the local grid.

Using energy with a lower carbon content

Our reliance on fossil fuels for energy is currently so profound that it will take many years to wean ourselves off it. But even within the family of fossil fuels, there are options that will lead to reduced carbon dioxide emissions. In brief: gas is better than oil, and oil is better than coal. The heat released in burning fossil fuels comes from breaking the chemical bond between the carbon and hydrogen atoms. Apart from soot and other nasties, the carbon (C) goes off as CO_2 (carbon dioxide) and the hydrogen (H) as H_2O (or water). So the greater the hydrogen-to-carbon ratio in the fuel, the less carbon dioxide is released for the same amount of energy. Thus, from a global warming point of view, it is better to burn methane (CH_4, the main constituent of natural gas, with four hydrogen atoms to every carbon atom), than it is to burn oil, which in turn is better than coal.

Ultimately it would be best to get our energy from renewable sources – wind, sun, and waves. Hydropower, biofuels, and nuclear power are sometimes lumped together with renewable energy sources, but because they carry particular risks we treat them separately below. Using renewable energy means tapping into nature's energy flows in "real time" and using it as it becomes available, not extracting stocks of ancient sunlight as is the case with fossil fuels. This immediacy puts us more closely in touch with the constraints of the global energy system as it is. It is unlikely to leave a horrible legacy for our children and grandchildren to deal with, and, best of all, it never runs out.

How much of our current and future requirements can renewable energy supply? George Monbiot makes a convincing argument for providing a significant portion of Britain's electricity needs with off-shore wind generators, with minimal innovation in long-distance transmission, which is critical, as windmills could then be sited a long way off shore where the wind howls and no one but sailors can see them. Countries with coastlines have the further option of using wave and tidal power. The technology is relatively simple, but has not yet been successfully implemented on a commercial scale. Different designs all make use of the basic principle that an object floating on the surface of the sea will be forced up and down because of waves and tides. This up and down motion of the float can be captured and transformed into electrical energy.

And what about the sun? In countries like South Africa and Australia – both very high emitters of carbon dioxide – it is crazy not to make use of energy from the sun. Paradoxically, both these countries have more coal than they know what to do with and export it all over the world. Because sunshine is available to all, it lends itself to household-scale installations like solar water heaters and photovoltaic panels. Both technologies are well researched, but remain marginal partly because of the high initial cost of the necessary equipment (after which your energy is free). Few governments or financial institutions provide incentives to make this investment attractive to individuals. Those with a tendency to conspiracy theories argue that research and investment in solar power has been sabotaged by governments and big business because of the technology's inherent ability to put power in the hands of people. There are a number of pilot plants that show it is technically feasible to generate solar electricity on a large scale. Research and development continue to make the technologies both more efficient and cheaper.

A few countries make some use of geothermal energy. In Iceland, Japan and the USA, steam from natural geysers is piped and fed to drive electric turbines. It is even possible to drill into hot rock, and pump in water to make a kind of artificial geyser. Unfortunately not all parts of the world have access to thermal energy sources.

Apart from the fact that many renewable technologies are outpriced by cheap fossil fuels, why do we not make more use of them? One problem is that energy densities are low. Generating 3,500 MW of electricity, roughly the capacity of a standard coal-fired power station, would take more than 2,000 windmills or over 60 square kilometres of solar photovoltaic panels. Other concerns have been raised about the ugliness of windmills (as if power stations were beautiful!) and their environmental impact, but none of these objections is insurmountable. One significant challenge with renewable energy sources is to maintain a reliable electricity supply because the sun does not always shine, and the wind does not always blow. With careful planning, however, and by optimizing the mix between technologies and their geographic spacing, most of these kinds of difficulties can be overcome.

Hydropower from large dams seems like a good option, but, in reality, it is not. In addition to the social and environmental costs of damming major rivers, the manufacture of the concrete used in their construction emits large quantities of carbon dioxide, and flooding valleys produces large amounts of methane. An alternative to dams are hanging turbines – a simple but as yet relatively untested concept of suspending self-contained turbines in fast-flowing rivers instead of damming them.

Biofuels, such as ethanol and biodiesel, have been put forward as a significant part of the solution to transport emissions. Replacing petrol and diesel with these renewable fuels will, according to their advocates, reduce carbon dioxide emissions. In reality, it is not quite so simple. Initially, using waste cooking oil from fast-food chains seemed like a great idea. Experimentally-minded environmentalists added caustic soda and other secret ingredients to old cooking oil to create diesel with which to fuel their cars. They felt justifiably smug that their vehicles were no longer consuming fossil fuels and contributing to climate change. But then the mainstream got hold of the idea. Biofuels are now big business and instead of waste products being used, vast tracts of land are being converted to grow fuel. Tropical forests are being chopped down for palm oil planta- tions which releases tonnes and tonnes of carbon dioxide that had been stored in the forest, not to mention the loss of biodiver- sity, livelihood, moisture, and so on that goes with destroying a rain forest. In other parts of the world maize and sugar cane are being grown as cash crops for ethanol instead of food. Pressure on agricultural lands in the South and the people that work them is likely to increase as Northern governments look for ways to replace petrol and diesel. Countries in the EU have recently committed to a ten per cent minimum target on the use of biofuels for transport by 2020. Environmental and development activists lobbied hard against this because at the scale envisaged, and the tropical rainforests that are likely to be cleared, these biofuels will in fact be a net producer of greenhouse gases.

The nuclear industry is also experiencing a resurgence as a non-fossil fuel source of electricity, but it is worth bearing in mind Amory Lovins's comment, made while working as a staff scientist with Friends of the Earth in the 1970s, that using nuclear reactions to boil water was about as efficient as "cutting butter with a chainsaw." Leaving aside the carbon dioxide emissions from the mining of uranium and its refinement, and the concrete used in construction, the technologies suffer from a range of other problems, and the risks are huge. No one has yet figured out how to dispose of highly radioactive waste that lasts for 100,000 years. The spectre of nuclear weapons' proliferation is never far behind and the risk of leaks and meltdowns is ever-present. Uranium, like coal and oil, is also a finite resource, and nuclear power generation is no cheaper than using renewables. Nuclear power is, at best, a last resort.

Carbon Capture and Storage (CCS)

What if, instead of reducing carbon emissions, we were to capture them? There is a nice logic to this, but there are several difficulties in putting this solution into practice. Here we look at capturing and storing carbon in forests, and, secondly, using technology to capture carbon dioxide at the point of emission, before burying it.

Plants are natural carbon sinks; they breathe in carbon dioxide and store it in their biomass as they grow. Trees are particularly good at this because of their size and longevity. The Kyoto Protocol (which you can read about in chapter eight) allows for emissions to be offset against new forests – carbon dioxide can continue to be emitted as long as trees are planted in proportion to these emissions. The problem with this is that it is hard to guarantee that a forest will be a permanent sink. Also not all trees are in forests. Some are grown in monocultural plantations that can badly disrupt the ecosystem they displace as well as suck up scarce water in dry areas. When trees die and rot, most of their carbon is released back into the atmosphere. Forests can burn down, or be felled, unless protected through laws and enforcement, which are subject to social and economic pressures. Natural forests continue to be denuded at an alarming rate for wood, and for land to grow food and cash crops. It is easy to see how a forest, grown as a sink under climate change law, could be chopped down as a result of such pressures.

Technically, it is feasible to "capture" the carbon dioxide that is released by coal- and oil-fired power stations, although full-scale viability may still be some years off. The captured carbon dioxide is compressed into a liquid and pumped deep underground. Leakages from these storage sites (old oil and gas fields, coal seams which cannot be mined, and saline aquifers) are possible, but minimal. In this way, carbon dioxide emissions from electricity generation in the UK could be reduced by 80 to 85 per cent, according to George Monbiot. But capturing carbon dioxide could also lead to *increased* exploitation of oil reserves because when carbon dioxide is stored underground in old oil reservoirs, it becomes possible to flush out the last of the oil that had previously been too expensive or too difficult to remove.

As with so many mitigation options, it is far easier and cheaper to build a power plant or cement kiln incorporating carbon capture than to retrofit an old one. Suitable storage sites have to be found close to coal seams or limestone quarries, otherwise transport emissions (and costs) will negate the whole project. Also CCS is only economically viable for large power stations, while there are other social, environmental, and economic arguments for building smaller stations. Apart from carbon dioxide emissions, there are other good social and environmental reasons to mine and burn less coal – acid rain, pollution and respiratory problems to name a few. CCS may be an option for coal-burning power stations and for cement manufacturing, but it is not really an option for dealing with emissions produced by oil. The products of oil refineries are used mainly to fuel cars, trucks, planes, and ships making capture and storage of emissions unfeasible.

The weird and the wacky

The enormity of the challenge presented by global warming has flushed out of the woodwork some strange people peddling some truly wacky ideas. It is well known that iron is one of the key elements that limits the growth of plankton and algae in the sea. One "scientist" has suggested that sprinkling millions of tonnes of iron filings into the sea will result in a plankton bloom large enough to soak up sufficient carbon dioxide to save the planet. All that would remain to be done would be to convince the little critters to sink to the sea bed and die before they release the carbon dioxide again. This experiment has actually been tried; it failed. Or what about capturing all the energy generated by people working out in city gyms? Hooking those treadmills up to the national electricity grid could save a fortune, as well letting gym-goers feel that they were really doing their bit for the planet. Or what about giant mirrors in space to deflect light and heat from the sun before it reaches us?

More seriously, millionaire businessman Richard Branson has issued a US$25 million challenge to anyone who can figure out how to remove a billion tonnes of carbon dioxide per year from the atmosphere. Given that his airline's fleet of jets contributes more to global warming than the whole of Ethiopia, for example, some might regard his offer as being long overdue.

We already have the technology to enable us to reduce emissions significantly in the near future. After that, as we struggle to reach and then maintain zero net emissions we will need all the innovative help we can get. So we should get our children to start dreaming up wonderful ideas to get us out of this mess. Hopefully one or more of these ideas will be implemented in our lifetime.

But if technology isn't the problem, at least for now, what is? And how do we tackle whatever that problem might be? We now need to explore the economics, politics, and philosophy of climate change, which we do in the next few chapters.

Chapter Seven

Why Is It So Hard to Mitigate Climate Change?

There are three main reasons why it is so hard to mitigate climate change: our dependence on fossil fuels (which has also created our wealth); the "tragedy of the commons"; and psychological inertia.

"Ruin is the destination toward which all men rush, each pursuing his own best interest in a society that believes in the freedom of the commons."
    Garrett Hardin, *The Tragedy of the Commons*, 1968

The fossil fuel economy

We have known about the link between carbon dioxide emissions and global warming for years, and the evidence keeps mounting up. So why do we continue to use fossil fuels? Why do we pump more and more carbon dioxide into the atmosphere? There is no simple answer to these questions, but it is clear that our individual lives, our households, and the economies of our countries are so deeply enmeshed with the use of fossil fuel that it is hard to imagine our world functioning in a different way.

Do you remember the last power failure you experienced? Perhaps it was during the summer and everything in your fridge went off. Or maybe you missed seeing a crucial penalty kick when your TV went blank. Or perhaps it happened in the middle of winter and, no matter how hard you tried, you just couldn't get warm. Maybe this is how you imagine a future world which has made the necessary adjustments to deal with climate change – one unending power cut.

That would not be an unreasonable assumption. Worldwide, public electricity supply and heating contribute almost 40 per cent of total carbon dioxide emissions. Electricity supplied to our homes has given us incredible options and kept us very comfortable. Yet, because most electricity is produced by coal-fired power stations, every time we switch on the computer, or the kettle, or run a hot bath, carbon dioxide is released into the atmosphere.

What is true at a household level is also true on a larger, industrial scale. Manufacturing, construction, and mining use vast amounts of energy, whether to produce food and the plastic and other containers it comes in, or the concrete blocks used to build luxury apartments, or the furniture inside them.

Before the industrial revolution, we relied on windmills, waterwheels, and the old-fashioned muscle-power of draught animals to provide the energy to keep the economy going. Now, most of the energy we use comes from fossil fuels, and it is the efficiency and high energy-density of fossil fuels that has allowed the unprecedented growth of some economies, and the accumulation of wealth. It would be fair to say that the reason industrialized, or developed, countries are rich is because they have consumed vast quantities of fossil fuel-energy and spewed great quantities of carbon dioxide into the atmosphere. But the use of fossil fuels is not the only problem; our patterns of land use are an often overlooked problem. Stable forests and grasslands around the world store billions of tonnes of carbon dioxide from the air. As more of this land is cleared to make way for settlements and farms, so significant amounts of carbon and methane are released back into the air. Feeding a growing global population is part of the pressure on land, but so is our insatiable appetite for meat. It is estimated that over 25,000 square kilometres of Amazon forest (about the size of Belgium) was cleared in 2004, mostly to grow soy as feed for cattle.

Besides agriculture, and the energy used in homes and factories, transport is the other major contributor to carbon dioxide emissions. Each time we drive a car or hop on a plane we add carbon dioxide to the air. And that little jar of imported jam, or punnet of "jet-fresh" vegetables, comes with a carbon dioxide tag. Trade between countries has risen exponentially over the past fifty years adding countless volumes of carbon dioxide as we exchange virtually identical goods – wine, cheese, clothes – over great distances, between countries equally capable of producing them. Often, a single item, like a car or a washing-machine, is assembled from many different components which have been made in different parts of the world, and shipped to be assembled where labour is cheap. And what about that take-away pizza? By the time it is in your hands its ingredients have effectively travelled thousands of "food miles". Why?

Economists are fond of a theory called "comparative advantage" which holds that it is more efficient for countries or regions to specialize in what they are good at producing and trade it for what they are not so good at producing. In an ideal world this should result in the best use of every resource: people, land, water, energy, and technology. But we don't live in an ideal world, of course. Firstly, global trade has been managed to suit the interests of the powerful countries which make the rules. Secondly, it has been managed without reflecting the true cost of transport. Thirdly, it has allowed developed countries to move towards cleaner, and less energy-intensive economies while still being able to import the products associated with smoke, pollution, and carbon dioxide from the less developed world.

"Climate change," according to the Stern Review (a report prepared for the UK government on the economic consequences of climate change), "is the greatest and widest-ranging market failure ever seen." The cost of cleaning up damage caused by climate change (which will have to be paid for by someone in the future) is not taken into account when fossil fuels are bought and used. Fossil fuels are the cheapest and most economically efficient form of energy for manufacture and transport only because we choose to ignore the costs of global warming, not to mention the costs to human and environmental health from other air and water pollution. This is a choice we have made in shaping our world economy; equally, we can choose to do things differently. Presented like this, the choice appears simple – what sort of fool would argue that it is better to keep burning fossil fuels and simply to take our chances with the consequences?

But there are obviously significant costs involved in shifting away from a fossil fuel-economy, and who will meet these? Developing countries locked into debt that they have repaid over and over again, still need foreign cash or hard currency. To get this they must manufacture and transport goods cheaply to sell to Europe, the United States, Canada, or Japan. For these developing countries, replacing cheap fossil fuels with some costly climate-friendly alternative is simply not an option. Closing down coal mines and refineries would put thousands out of work, and it wouldn't help them to know that their sacrifice had helped some other "cleaner" sector of the economy to grow, or that the planet was now less likely to overheat. Simply taxing fossil fuels would make goods more expensive and put them out of reach of many poor people. What is required is a *just* transition to a post-oil economy. Otherwise the cost of transforming societies and economies will be borne by the poor and powerless, rather than shared in a way that is equitable and practicable.

There are some economists who like to compare the cost of doing something about global warming with the cost of doing nothing. This seems an unhelpful and simplistic approach, though. The question is not, "Is it more expensive to mitigate or to adapt?" but, rather, "How, knowing what we do about the sensitivity of the planet to greenhouse gases, should we be shaping our economies to meet people's needs?"

The tragedy of the commons
Given that the individual economies of the world are so dependent on fossil fuels and on a global economy, finding a relatively just and painless way to de-couple economic production from carbon dioxide emissions will require global coopera-

tion. As things stand, continuing to burn fossil fuel brings uneven benefits to the economies of individual countries, while spreading its consequences to everyone in the world (and not even very equally, as we show in chapter four). Sadly, global politics seems to be conducted mostly on the basis of "whoever blinks first, loses". Read between the lines of the reports emanating from any international conference on climate change and you will see politicians from one country refusing to act to combat climate change unless another country (and here they usually name the USA or China) acts first. In the end no one benefits.

In his famous 1968 essay *The Tragedy of the Commons*, Garrett Hardin describes the overgrazing of a piece of communal land, a sobering analogy to climate change.

> The tragedy of the commons develops in this way. Picture a pasture open to all. It is to be expected that each herdsman will try to keep as many cattle as possible on the commons. Such an arrangement may work reasonably satisfactorily for centuries because tribal wars, poaching, and disease keep the numbers of both man and beast well below the carrying capacity of the land. Finally, however, comes the day of reckoning, that is, the day when the long-desired goal of social stability becomes a reality. At this point, the inherent logic of the commons remorselessly generates tragedy.

Hardin goes on to describe the "inherent logic" by which each herdsman will add one more animal to graze on the common pasture because they get the direct benefit, whereas the cost of an over-grazed pasture is spread among everyone. Each person adds relentlessly to this situation and, as Hardin puts it, "Therein is the tragedy."

Climate change presents the same problem, and we have reached our "day of reckoning". The air is a common resource with open access. Without agreed rules, it is in an individual business or country's (short-term) interest to continue to spew carbon dioxide, methane, and all the other greenhouse gases into the air. And the consequences of global warming are shared by all, even those who contribute no greenhouse gas emissions at all.

The way around the tragedy of the commons is through co-operation. In chapter eight, we will look at what co-operation there has been, and why it has been so slow to come about.

Psychological inertia

There is no doubt that most people respond quickly and with a great deal of compassion in an humanitarian crisis. Think how fast funds were raised around the world from ordinary women and men to respond to the devastating impacts of the tsunami that hit on 26 December 2004. People saw what was happening, how it was affecting others and believed that they could do something to help. So they did. Global warming isn't like that; it creeps up on us slowly and relentlessly, making the lives of so many people who are already struggling to survive, yet more miserable. It is possible that global warming will result in some monumental and devastating event, like the melting of the polar ice caps, but we need to act well before this becomes a likelihood. Although this may be hard for us to grasp, we have to begin *now* to take significant steps to combat global warming.

Our being so wedded to fossil fuel-economies and the fact that the air is a "global common" offer a partial explanation for our inaction. But there is also a kind of psychological inertia which feeds our collective apathy and explains the myriad reasons why people – us, politicians, heads of large corporations – simply do nothing.

Part of the problem is an inability to associate our actions (e.g., driving a car, using disposable plastic packaging) with their impacts (e.g., sea level rise threatening the existence of a small Pacific island). The impacts are often far away and delayed, and it does not seem clear that we have caused them. Think of military personnel operating what seem to be video games, but in fact depict real people, or cities in which real people live. Simply pressing a button launches a missile, and the screen shows that the target has been hit. Any of us would find this psychologically easier than killing a sleeping civilian in cold blood with our bare hands.

Another aspect of the problem is a sense of impotence. One

person's commitment to fighting global warming might include a decision to walk to the shop, rather than driving, or to replace an electrical boiler with a solar-heated water system, but, in reality, either or both of these actions would have a negligible impact on global warming. Given the scale of the problem, we might wonder if there really is anything meaningful one can do (see chapter ten if you want to skip the philosophy and go straight to some practical suggestions).

Another part of the problem is that most of us don't like being told what to do. What is their motivation? What are their vested interests? What about my personal freedom? These are important questions, but they can prevent us from taking any action at all. We don't always trust our own politicians and scientists, let alone the political leaders of other countries.

And so it goes; our disbelief, impotence, suspicion, and laziness all conspire to entrench a psychological inertia, but by recognizing this, as a first step, we can begin to turn things around.

Chapter Eight

Co-operation and the Corridors of Power

This chapter explores what is fair, and why fairness matters. It describes what goes on in the corridors of power and how different countries use their particular histories and current realities to enter the negotiation game. It also provides a brief history of the FCCC, explains what the Kyoto Protocol is, and how its targets compare with what is needed and what is fair.

Esso gave more than any other oil company towards the 2004 US elections, the overwhelming majority of it to the Republicans. One of George Bush's first actions on being elected was to pull the USA out of the Kyoto Protocol, just what Esso had been campaigning for. Esso has also spent more than $12 million since 1997 funding think-tanks and lobby groups, such as the Competitive Enterprise Institute which argues that climate change will create a "milder, greener, more prosperous world" and that climate change is a myth put about by the EU to hamper US competitiveness. As Greenpeace puts it, Esso chooses to wreck the climate; we can choose not to buy Esso products.

But what else is going on in the corridors of power of big business and government? Why have we left governments – notorious for acting in their own interests, to protect local industries, for example – to solve such an important global problem? More to the point, what have they actually done? To answer these questions we will explore briefly the history of international co-operation and look at what has been agreed. We will look at how governments work together – or not – and how they use their particular national histories and lobby groups in a game that attempts to assign responsibilities and tasks for cleaning up the mess.

Genesis of the FCCC and the Kyoto Protocol

In 1979, almost thirty years ago, governments of the world gathered round at a United Nations conference and looked at what the scientists had to say. Global warming was a problem, they agreed. But they failed to agree on any steps that should be taken to address it.

In the 1980s, global warming was put on the back-burner while the world focused on the ozone hole and how to eliminate the chlorofluorocarbons (CFCs) that were causing it. Ozone (a relatively unstable compound of three oxygen molecules) acts to screen out most of the sun's harmful ultraviolet rays, and CFCs (man-made gases, initially seen as an environmental blessing because they are so stable) were making their way into the stratosphere and destroying ozone molecules. Life without the ozone layer would be impossible (with the possible exception of deep under water). So the world's governments acted fairly swiftly and agreed to phase out ozone-depleting substances. In terms of the Vienna Convention and its Montreal Protocol, the principle of "common but differentiated responsibility" was successfully applied. It was the responsibility of all of us to take action, but those actions would be different for each country. Developed countries had to act immediately to stop producing CFCs (and other ozone-depleting gases) while developing countries had a ten-year grace period in which to comply. In many respects this was a similar issue, if more straightforward, than tackling global warming, and a practice round of global co-operation.

The 1980s were also the decade of the global warming-deniers. Their voices can still be heard to this day, though few bother to listen any more. Spurred on by the tobacco industry's success in casting doubt on the link between smoking and cancer, the global warming-deniers took on global warming with a two-pronged strategy. Their first tactic was to create confusion in the public mind about the scientific evidence (we hope that the first part of this book has given you ample evidence that climate change is a real concern, that there is scientific consensus, and that there is a fundamental difference between uncertainty and a lack of consensus). Their second tactic was to pretend that they are "neutral", without a vested interest. Many deniers, although they may operate as civic groups or non-profit organizations with nice, green-sounding names, have been shown to have close financial and other links to the oil and coal industries. Their letters and press statements, which are couched in pseudo-scientific language, are more often than not, signed by some "eminent scientist" (with just the right number of initials and acronyms behind his or her name), and make references to obscure research papers. The media, always on the lookout for controversy and strong opinions, give them a far higher public profile than they deserve. Global warming-deniers have done a lot to impede co-operation and to manipulate public perception, and have thereby seriously delayed an effective political response to global warming over the last twenty or thirty years. Our children would be justified in looking back on this delay with incomprehension at our greed and ignorance.

Nevertheless, some significant progress was made. In 1988, the World Meteorological Organisation (WMO) and the United Nations Environment Programme (UNEP) set up an important multi-country institution called the Intergovernmental Panel on Climate Change (IPCC) to assess scientific information related to various components of climate change and to formulate realistic response strategies. (An interesting aspect of the IPCC is that, although it seeks to draw conclusions from the research of hundreds of scientists, its summary reports for policy-makers are not issued without the approval of national political representatives. Thus there is a tendency for the reports to be toned down to what Flannery refers to as "lowest common denominator science.")

In 1990, the IPCC issued its first report, and, in 1992, more than a hundred heads of state, thousands of government representatives and tens of thousands of ordinary concerned people met in Rio de Janeiro, Brazil, to discuss the state of the planet. They made speeches and commitments, signed statements and treaties, and agreed that we all had to do things differently. This was an important first step. One of the documents they signed was the United Nations Framework Convention on Climate Change (FCCC) which recognized the problem of global warming at the highest level, and contained an agreement on the need to *stabilize* concentrations of greenhouse gases in the atmosphere. It contained no specific targets (which had proved too much of a political hot potato), either for limiting the quantity of greenhouse gases emitted, or any indication of what would constitute a dangerous level of greenhouse gases in the atmosphere. What it *did* include was a commitment to set targets at some point in the future – first, for developed countries and economies in transition, which were listed in Annex 1 to the treaty. In international climate-speak, developing countries are referred to as Non-Annex 1 countries. The FCCC also included a commitment to both the principle of a common but differentiated responsibility and the precautionary principle (to err on the side of caution, even if the scale and the magnitude of the risk are not fully understood – a bit like keeping a suspected multiple murderer locked up even though he has not yet been found guilty). To set the whole machinery in motion, it was also agreed to establish a Secretariat to organize more meetings, a body to provide technical and scientific advice, and another to provide funding. Countries are required to report regularly on their greenhouse gas emissions, within agreed accounting guidelines.

In December 1997, five years after Rio, the third Conference of Parties to the FCCC (a meeting of countries who had signed the treaty and more commonly known as COP) met in Japan to agree on targets to reduce emissions for six greenhouse gases.

These countries signed the Kyoto Protocol, but, as we will see, not all of them would stay the course. Kyoto agreed that Annex 1 countries would bring their emission levels down to at least 5 per cent below 1990 levels during the commitment period, 2008 to 2012. Not exactly radical, and not based on science, but it was a start and it was seen to be achievable. Some countries managed to argue for less stringent targets, and some, like Australia, were even allowed to *increase* their emissions! Such are the mysteries of global politics.

Countries agreed that they should have several ways to meet their targets. The most preferred, and the least controversial, would be through direct reduction of their greenhouse gases emissions, by using smaller quantities of fossil fuels, and using them more efficiently. Secondly, targets could be reached by increasing the capacity of carbon sinks, such as new forests. Thirdly, countries could use the three "flexible mechanisms" of joint implementation, emissions trading and the clean development mechanism (these so-called "flex-mechs" are discussed in more detail in chapter nine). In brief, they are ways in which countries can avoid the difficulties of changing things at home, which is why they are called flexible. The net global result should be the same, but cheaper to achieve. Desperate for an easy way out, countries could not agree on how much of their targets could be reached through these mechanisms, and, legally, a country could offset its entire target if it could afford it. But the intention, as stated in the Protocol, was that the flex-mechs would be "supplemental to domestic actions." At a later COP, in Marrakech in 2001, it was further agreed that domestic efforts, as opposed to mechanisms, should (in the wonderfully understated language of international negotiations) constitute a "significant element". The Marrakech Accords also allowed for businesses and NGOs to participate in the three mechanisms, under the responsibility of their respective governments.

Although signed in 1997, the Kyoto Protocol would only enter into force when ratified by at least fifty-five of the signatory countries, which together accounted for at least 55 per cent of all emissions from Annex 1 countries. The commitment of the big emitters like the EU, Russia, Japan, and USA (which produces almost a quarter of all the carbon dioxide pumped into the atmosphere) was obviously critical. "The American lifestyle is not negotiable," (or words to that effect) said George Bush Senior, son Dubya, and key advisor Dick Cheney during various stages of the climate change discussion. In other words, "Don't expect us to drive smaller cars." Ten years after Kyoto this remains the USA's position (although not the position of many US cities which have put their names behind the Kyoto Protocol). In the end, it took until 16 February 2005 – more than seven years – for the Kyoto Protocol to enter into force, when, at last, Russia, keen to join the World Trade Organisation, or so rumour has it, finally signed on the dotted line.

Hell no, we won't sign!

The USA has provided two reasons for their refusal to ratify the Kyoto Protocol. They say it will hurt US business, but the same argument could be made for every country and is precisely the reason why it is so important for all major emitters to make joint commitments. In that way, any harm that may be done is equal, and the *relative* harm is minimal. The second reason given by US negotiators is that they will make no commitments until developing countries also make commitments, implying that that would be the only fair way to proceed. But it is not. No matter which way one looks at it – in terms of total annual emissions, per capita emissions, or historical cumulative emissions – the USA is by far the biggest carbon dioxide polluter. Fairness, in any sense of the word, would be for the USA to lead the clean-up process, even if they weren't also wealthy enough to actually do this.

There is a third, unspoken, reason why the USA remains outside the climate change negotiations. Traditionally, the US government will not agree to any international commitments that are not already part of domestic regulation. They don't like the rest of the world telling them what to do, and, unsurprisingly, the rest of the world doesn't like to be told what to do by the USA either. But despite their reluctance to be party to international agreements, Americans are taking climate change seriously, spending much more money on climate change research than any other country. Several states, large cities, and business associations have adopted self-imposed emission and renewable energy targets.

Australia is the other big emitter that has not signed the Kyoto Protocol, even though they managed to negotiate a Kyoto target that was *higher* than their baseline year and can thus emit *more* greenhouse gases than they did in 1990. This is very odd, as their per capita carbon dioxide emissions in 2000 were 17.4 tonnes, the fourth highest per person in the world.

By contrast, for the year 2000, India and China (while they are significant emitters) produced only 1 and 2.7 tonnes per person of carbon dioxide respectively. During the same year, the USA emitted a whopping 20.2 tonnes per person. If one takes cumulative emissions from fossil fuel and cement production in the fifty years between 1950 and 2000, sub-Saharan Africa generated a mere 13,867 million tonnes (almost three-quarters of which came from South Africa) compared to 229,327 million tonnes generated by North America (more than sixteen times as much), and 292,323 million tonnes generated by Europe (twenty-one times as much). Emissions from changes in land use are relatively higher from Africa partly because, by 1950, much of Europe and North America had already been deforested. In fact, in the USA there was a net sink during those fifty years – more carbon dioxide was captured than was released by changes in land use as a result of forests growing back where they had been denuded.

The low emissions from developing countries do not mean that they should not look seriously at their development paths, simply that, to date, the contribution of each person in the South to global warming is minuscule compared to that of each person in the North. How do we know this?

Counting carbon

There are different ways of measuring how much a person or country has contributed to global warming. Countries like to pick the ones that make them look good, or show up other countries as global warming villains. For example, one can compare total current emissions of carbon dioxide (the total amount spewed into the air by each country over a period of, say, one year). This makes the United States look particularly bad as it contributes almost a quarter of total carbon dioxide emissions. By this yardstick, China and Japan don't look so good either. One can also compare current emissions of *all* greenhouse gases by converting them into a carbon dioxide equivalent. By this measure, the Northern countries still look bad, but a little less so, because the countries of the South are responsible for a lot of the methane released by agriculture, notably rice farming.

One can take either of these numbers and divide them by the number of people living in those countries to get per capita emissions. Countries like Nepal and Mozambique hardly register, while Canada and the USA are the real villains. Or one can divide the numbers by the country's Gross Domestic Product to get the economy's "carbon intensity". The USA looks marginally better when this indicator is used, and one of George W. Bush's nominal commitments to combating climate change is to reduce the greenhouse gas *intensity* of the US economy by 18 per cent over the next decade. This may give him something to say, and make him feel good, but it does very little to combat global warming. In real terms, this means that although fewer greenhouse gases are emitted per dollar of GDP (which is not a bad thing), growth in GDP results in an overall *increase* in carbon dioxide pumped into the atmosphere.

And then, to really complicate things, by incorporating issues of equity and impact severity, you can look to the past or the future. The reason the planet is warming is because of *cumulative* emissions. It is not because we pumped 26.4 gigatonnes of carbon dioxide per year into the atmosphere over the previous five years, but because we have been emitting larger and larger quantities of greenhouse gases for the past two hundred and fifty years. It is this accumulation of carbon dioxide in the atmosphere that causes global warming. When you think of it in this way, you can see why developing countries make the important point that their development should not be penalized because of the problem caused by countries that have already industrialised.

Even when looking at only the second half of the last century (1950–2000), it is clear that developing countries have contributed very little to global warming. The USA and Europe between them contributed over half of cumulative emissions (27 per cent and 24 per cent respectively). Developed countries accounted for more than three quarters of emissions from fossil fuel and cement; developing countries, comprising far more land and people, contributed less than a quarter. When changes in land use are taken into account, cumulative emissions from developing countries are more significant, but remember that this is because developed countries cleared most of their land of forests long before 1950.

	Cumulative CO_2 emissions 1950–2000					
	Fossil fuel and cement % Total		*Changes in land use* %Total		*% Total*	
Developed	76.2	598,135	0.2	655	54.6	598,790
Developing	23.8	186,721	99.8	310,586	45.4	497,406
Total		784,856		311,241		1,096,196

Source: World Resources Institute. CO_2 emissions given in gigatonnes.

Political negotiations

While global politics may display a veneer of diplomacy and civility, beneath it all, might is right. Whether it is fair or not, a lot of effort goes into protecting privilege and maintaining the status quo. Thus, those who are currently emitting the most carbon dioxide assume the *de facto* right to continue emitting – possession is nine-tenths of the law – and any mitigation is based on this "current right". It is not surprising to learn that those who own this right to emit are also the richest and most powerful countries. It is they who set the rules.

The corridors of power are filled with people trying to get the best deals for their countries. Each country that is not in complete denial about climate change wants two, conflicting, things. Firstly, they want *global* targets to reduce emissions to be as *high* as possible. This would minimize the impact on climate change and, thus, the costs of adaptation. Secondly, each country's own *national* target should be as *low* as possible. This would minimize the costs to the negotiator's country of mitigating climate change. Or – translated into the language of *The Tragedy of the Commons* – each cattle owner wants the size of the total herd (greenhouse gas emissions) to be below the carrying capacity of the pasture, but he wants most of the cows to be his. It is readily apparent that the targets for cutting greenhouse gases (a mere 5 per cent under Kyoto) bear little relation to scientific requirements. Instead, these targets reflect a "political reality" as each country, keen to maintain its own emissions levels, sacrifices the global goal.

In fact, this is not quite accurate. Some countries – those producing oil, for example – have little interest in constraining the use of fossil fuels globally. Fossil fuels are the source of their revenue and most of them are unlikely to be among the hardest hit by climate change. Small island states, on the other hand, will do almost anything to reduce global emissions because a rise in sea level could leave their countries under water. As fate would have it, Saudi Arabia and Tuvalu are both members of the G77, a negotiating bloc established in 1964 to strengthen developing countries' participation in international negotiations. It now has over 100 members and is a diverse collection of countries with sometimes contradictory goals. On an issue such as climate change, there are clear tensions within it.

The position of the G8 (the eight richest countries in the world) is also racked with internal tension. The USA, having thrown its toys out of the cot, has climbed right out and is going it alone. Despite not being a Kyoto signatory, the USA did sign the FCCC and so continues to participate in some of the ongoing negotiations. The EU, itself a product of consensus politics, is well practised in these kinds of negotiations and keen to keep the FCCC moving ahead. To do this they have made an interesting opening bid for the next round of negotiations under the Kyoto Protocol. They will reduce greenhouse gas emissions by 2020 to 20 per cent below 1990 levels regardless of what anyone else does. They will up this target to 30 per cent if an agreement can be reached with the USA and other key countries. A tasty piece of bait for the world's highest polluter.

And now?

A quick look at 2003 figures shows that while some Kyoto signatories, including about a third of the EU members, are well placed to meet their Kyoto targets in the 2008 to 2012 period, others, most notably Canada and New Zealand, have *increased* their emissions significantly since 1990. Emissions from economies in transition have dropped significantly and to well below their Kyoto targets of 8 per cent. Emissions from the Baltic states have dropped by extraordinary amounts; by between 65.9 per cent (Estonia) and 77.5 per cent (Lithuania). Although these great reductions in emissions are due to economies being in freefall, rather than to any targeted strategy to combat climate change, they are of the magnitude that is required according to the science. It is likely that the next round of Kyoto commitments will be much more stringent. Conversations in the corridors of the climate change negotiations concern two broad topics: firstly, what new targets will Annex 1 countries set; and secondly, how to bring the USA back on board, and how to set targets for developing countries. (There is a growing recognition that to get the USA to sign, which is essential, will require at least the better off developing countries to agree to emission targets.)

The existence of the FCCC and the Kyoto Protocol indicate that there is global co-operation of a sort. Although the Kyoto commitments to reduce emissions are wholly inadequate, the commitment of most countries to work together is important. So is the fact that the Protocol did not collapse when the USA withdrew in 2001. Negotiations have already started for a post-Kyoto agreement. Three things will indicate that progress is being made. One – agreement on what is a dangerous level of greenhouse gases in the atmosphere and a global emissions cap. Two – re-entry of the USA into global commitments. Three – recognition and further elaboration of common but differentiated responsibilities.

Chapter Nine

Flex Mechs, Offsets, and Other Mitigation Measures

This chapter shows that although it is difficult to mitigate global warming, we have no choice. The challenge is to find the least painful and fairest way to meet tough targets. Mitigating climate change provides an opportunity to address some of the worst aspects of our globalized economy, including the growing divide between rich and poor – there is no reason for thousands to die needlessly each day in a world which is so rich in resources. For developing countries, the challenge is to find a way to create sufficient wealth to meet people's needs without relying on fossil fuel energy.

"Buying and selling carbon offsets is like pushing the food around on your plate to create the impression that you have eaten it."

George Monbiot

The challenge to reduce our dependence on fossil fuels and to stop destroying forests is both daunting and exciting. It is daunting because the task is so immense, but exciting, too, because it is an opportunity to fundamentally transform the global economy. There is a lot that we can do, as individuals (see the next chapter), but to make an impact on a problem of this magnitude requires concerted action on many levels and right across the world. Policy and practice at international, national, local, and individual level must be co-ordinated and effective. Some of the excitement of the challenge lies in the opportunity to get rid of what is wrong with our global economy, while retaining what works. Would the world really be a poorer place if cut-flowers and tobacco were no longer grown for export in countries where farmers should really be growing food, for their own consumption?

It is tempting to argue that stopping global warming is so important that it does not matter how we do it, that the end justifies the means. History has shown time and time again that this belief – no matter how well intentioned – has led to extreme human misery, including genocide. There is no reason to believe that tackling global warming will be any different. So *how* we do things *does* matter; the means determines the end. If we are to build a world in which everyone has enough to eat, and a chance to grow and learn, rather than one which entrenches inequalities and fear, we need to choose with care how we mitigate global warming. It all starts with how we allocate the right to emit.

Contraction and convergence

Aubrey Meyer, an English concert viola player among other things, has proposed the concept of "contraction and convergence" as a reasonably fair way to allocate and cut carbon dioxide emissions. In brief, one decides how much carbon dioxide the planet can safely absorb. This number is divided by the number of people on the planet so that everyone receives the same right to emit carbon dioxide (or greenhouse gases). A country's target is simply this right to emit multiplied by the population, to be met within a set time frame. Some countries – in North America, Europe, etc. – would need to reduce their emissions substantially, or contract, while others – Mozambique and Honduras for example – could increase their emissions until they converged with the global norm.

This method is not without its problems. First the world would have to agree on the maximum safe amount of carbon dioxide and other greenhouse gases the air can absorb. This is not as easy as it seems (see chapter eight), and the figure arrived at would probably be a compromise between the science and political viability. Then the world would have to agree on a baseline year for counting populations – otherwise there might be an incentive for a population explosion. Agreement would also have to be reached on a time frame for meeting the targets. Finally, it would have to be decided whether or not historical emissions counted, and to what degree – perhaps the trickiest issue of all.

If the quantity of greenhouse gases that the earth can reasonably accommodate is represented as a giant pizza, more than half of that pizza has already been eaten, and by only a quarter of the world's population. This represents the historical emissions of rich countries. Now that everyone wants a share, how do we divide the remainder? Is it fair that those who ate so much in the past should continue to eat as much? On the other hand, is it fair to penalize them for eating as much as they did in the past when they thought that they could order as many pizzas as they wanted?

Limiting carbon dioxide emissions without accounting for historical emissions means that Southern countries will not be able to build their wealth by burning fossil fuels, in the way that their Northern counterparts did. Will they be prepared to make this sacrifice? Would the result be eternal dependence on handouts? On the other hand, taking historical emissions into account means penalizing Northern countries for doing something they did not know was a problem when they did it. But one does not need to account for all of history and some pragmatic compromise is possible. The clock could be started in 1990, the baseline year for the Kyoto Protocol. By then, everyone knew of the link between carbon dioxide emissions and global warming. This baseline still penalizes Southern countries and so, to be fair, compensatory measures need to be built in to any agreement. Writing off the South's financial debt against the ecological debt owed by the North to the South could be one such measure. A quick calculation on the back of an envelope shows that the industrialized countries would be getting off lightly, but still, it would be something. And with these details carefully negotiated, the model of contraction and convergence would, on the whole, be a fair and equitable way of distributing the benefits of fossil fuels and the costs of global warming.

An alternative, or additional, way to compensate for historical emissions would be to turn technology transfer on its head. Instead of technology transfer (along with annoying advisors who disgorge often unhelpful advice along with tonnes of carbon dioxide as they fly to and fro) being seen as a favour done by the North to the South, it could become something that developing countries demand as their right. And the beneficiaries could decide what form the technology transfer should take.

The contraction and convergence model provides another more fundamental mechanism for wealth transfer that could help shift the balance of power in the world. If such a model were negotiated, the developed world (those that ate most of the pizza and still have bloated appetites) would have to buy or borrow emission rights from developing countries. These now poor countries would become rich (because they would have so much carbon dioxide to sell) and the rich countries would become poor because their carbon debt is so huge. No longer would developing countries need to orient their economies to meet the needs of people in the rich world, but people in the rich world would need to orient their economies to meet the needs of people in the poor world! At best and with the right supplementary agreement and regulation, this could mean that instead of producing things for people who consume the most (and are the richest and most wasteful), we would produce things for people who consume the least. People who currently emit negligible greenhouse gases would have gained purchasing power.

Rationing carbon

Carbon rationing takes the contraction and convergence model a step further. An annual carbon ration could be assigned for each person (the total amount of carbon dioxide directly or indirectly emitted by each person would be limited to this rationed amount). Everything we consume has a carbon content and it would be possible, if very tricky, to pay both a monetary and a carbon price for each thing we buy. It would be simpler to limit the rations only for electricity and fuel, and, perhaps, items involving large amounts of concrete. Because companies would also be allocated (or would buy) a limited carbon allowance, the price of each good would quickly reflect its carbon content. Rations could be traded, and those who wanted to emit more than their fair share of greenhouse gases would have to buy extra rations. Some limits would have to be enforced, though; it would be untenable if people were driven to sell the bare minimum needed to cook food and stay warm for cash. Other mechanisms would also be needed, to ensure that carbon rationing didn't act as a perverse tax on poor people. For example, while it would be relatively easy for a wealthy family to replace its boiler with a solar water-heating system (thereby decreasing their domestic carbon dioxide emissions by up to 40 per cent), in most countries that step would be too expensive for a poor family to take.

Government policy regulation, incentives, and taxes
What *we* can do to limit *our* carbon contribution will be dealt with in the next chapter. But in the very near future, our actions will be, or should be, taking place within a framework of policies and laws to support such actions, and to prevent abuse and perverse effects. Other laws may be required to force business, industry, and individuals to reduce their greenhouse gas emissions.

Carbon rationing and other economic mechanisms would have to be well governed and supported by a range of legislative measures, and not only to prevent abuse. It would be difficult for an individual to stay within his or her carbon ration while living in a contemporary Western-style economy. This would create a lot of legitimate pressure to increase the carbon budget, so something more will have to be done.

Transport is one area in which government intervention is essential. Public transport improvements and a moratorium on six-lane motorways would encourage more commuters out of their cars. At an international level, the situation is more complex. Which government is responsible for regulating the emissions from a ship registered, for example, in Panama, which stopped in Montevideo to re-fuel (with Saudi oil) and sailed on to Lagos with a cargo of rice from Thailand – and burned up tonnes of oil in the process? Air travel is marginally less complex, but will also need to be dealt with through international treaties. (Note that neither ships nor planes that cross national borders are included in the Kyoto Protocol's current targets). Perhaps, in the meantime, governments could place a moratorium on new or expanded airports.

Using the same amount of energy that we use now, but more efficiently, could save tonnes of carbon dioxide from being emitted and a great deal of money too. Yet energy-efficiency remains the lost child of anti-global warming strategies. Perhaps this is because increased energy efficiency is not as sexy as carbon capture and other big engineering solutions. No one has ever been awarded the Nobel Prize for efficiency! Perhaps it is also because electricity is, relatively, so cheap. It may also be because those little pamphlets on how to save energy are often published by power utilities – and they're the ones that make money selling it to you in the first place. It's no wonder that messages get mixed.

One way to encourage the efficient use of electricity would be to increase its price, although industry could simply pass on the higher cost, or most of it, to consumers. That mechanism would have to be combined with in-house efficiency programmes. One study in South Africa found, for example, that, by making fairly simple changes, a car-manufacturing plant could save 16 per cent of its energy costs with a pay-back period of one year. Implementing only partial measures could save them two million rand (about US$300,000) per year with a one month pay-back period. But the plant had not made any of the recommended changes. And why not? The production manager did not have energy efficiency as one of his performance measures.

There are many other steps that governments could take. To make an informed "global warming-sensitive" choice about a new washing machine, one needs to know how much electricity it consumes per wash-load, compared to its competitors –

governments could make such labelling mandatory. In some countries and states, consumers have the choice of buying "green" electricity, at a small premium, from wind or other renewable sources. This is a way for consumers to send a signal to power producers about which forms of power production they prefer, and so encourage the "green power" market to grow.

National and local governments could offer tax incentives for installing photovoltaic (PV) panels and solar water heaters in businesses and households, or they could set up and manage financing schemes to enable people to instal these. The loans could be "soft", repaid in the form of energy savings rather than as cash. In some instances, where buildings are equipped with PV panels, any excess electricity is fed back into the local electricity grid. The electricity meter then runs backwards and earns the occupants credit. Incentives may help to ease these technologies into the marketplace and to make them increasingly viable. Ultimately, though, governments will have to change building codes to make it mandatory for appropriate energy efficiency to be designed into all new buildings.

Carbon off-setting

Most responses to global warming require us to change our behaviour in some way, and that can be a drag. Carbon offsetting is a desperate attempt to avoid this – a mechanism that lets you fly or drive to your favourite holiday destination or the next climate change conference with a clear conscience. Sound too good to be true? It is. Carbon offsetting is a philosophy that says it's okay to emit three tonnes of carbon dioxide on Monday as long as I take steps on Tuesday to capture or offset an equivalent amount. I can plant a tree (the usual favourite) which in its lifetime will capture and store about 3 tonnes of carbon dioxide. I can invest in a solar water heater which will save 3 tonnes of carbon dioxide that would have gone up the chimney. Mostly, offsetting seems a way to assuage guilt, that scourge of Western minds. Sorry to be a killjoy, but if we're going to save the planet we need not only to reduce (or better yet, eliminate) our flights and road trips but also to support projects that generate renewable energy, or replant forests in denuded areas. Offsetting doesn't really get around the fact that the last jetliner you flew in put an awful lot of carbon dioxide into the air.

Offsetting can be used to ease our personal guilt, but it can also operate at a collective level. Under the Kyoto Protocol there are two ways in which a country can legally (though perhaps not in reality) offset its emissions. The first is through carbon sinks. Each FCCC signatory country has to provide annual carbon accounts. On the "plus side" are all the greenhouse gases that have been pumped into the air, while on the "minus side" is the carbon dioxide calculated to have been absorbed by the new forests (or plantations) that have been planted (though not the existing forests that might have been protected). So, if you plant a forest, you can carry on emitting.

The "Flex Mechs" Joint Implementation and Emissions Trading have been discussed above, so we look now at the Clean Development Mechanism (CDM). Under this, you can support a carbon-saving project in a non-Annex 1 country (i.e., Southern countries which have no commitment targets under Kyoto). Instead of reducing emissions in one's own country, one can reduce either present or future emissions in *another* country! Sounds great! The criteria for projects is that they should support sustainable development and they should be "additional" and not something that the target country would have done anyway. Both sinks (forests) and emission savings can be considered under the CDM. The CDM is not, in fact, a bad idea in principle. It could, in theory, provide much needed resources to help developing countries to shift their development paths to models based on renewable energy.

CDM – even its offsetting component – could work if operating within a tougher system of contraction and convergence (i.e., if the targets set for all countries were significant enough to make us pause as we hurtle towards climate disaster). Like emissions trading, CDM could be a way of implementing anti-climate change measures in a more cost-effective way, while supporting countries to make climate-friendly choices when it comes to development options. As it is, it seems crazy that Annex 1 countries can use the smokescreen of CDM projects to obscure the extremely limited cuts they have made in their own emissions.

Emissions trading and joint implementation

No matter what the item on the agenda, national delegations to international treaty negotiations are seldom without a strong trade representative. Little wonder, then, that emissions trading was strongly promoted as a mechanism to solve global warming. It is one of three flexible mechanisms (flex mechs) under the Kyoto Protocol, but only available to Annex 1 countries. And there is a precedent for its succesful implementation. Sulphur dioxide – which results in acid rain – was effectively curtailed in parts of the USA through setting an emissions cap and introducing a trading system. Europe has recently initiated a carbon dioxide emissions trading system.

Countries, or companies, which struggle to meet their emission reduction targets can buy units of "hot air" from countries, or companies, which exceed theirs. The end result as far as global warming is concerned is the same, only, according to conventional economics, cheaper. The logic is as follows: turning the right to emit carbon into a commodity that can be traded subjects it to the laws of supply and demand, as with any other commodity. If it costs more to reduce emissions by one tonne than to buy the right to emit one tonne, countries and companies will buy that right. That pushes up the price of "hot air" until, at some point, it becomes cheaper to reduce emissions.

Emissions trading in itself does not encourage reduced emissions of carbon dioxide. That is done by setting a cap: a figure that is scientifically meaningful, politically acceptable, and technically achievable. Many activists argue that these caps are always too high, and effectively give countries, or companies, the right to pollute. In other words, there are too many units of "hot air" on sale. There are also concerns about accounting accuracy. But some environmental organizations are already participating. They raise money from their supporters to buy up these hot air units, but don't use them. This effectively lowers the cap and, thus, overall carbon dioxide emissions.

Joint implementation (JI) is another mechanism recognized by the Kyoto Protocol as a way of reducing costs and reaching targets. Under the JI, any two countries which each have a target to meet can co-operate on a joint project and share whatever carbon credits arise. For example, Sweden can part-finance the installation of thousands of solar water heaters in Greece. From a global warming perspective, a certain amount of carbon dioxide is saved. But from a financial point of view, since Greece is sunnier than Sweden, you get more carbon dioxide saving for your money. The JI mechanism sets rules for exactly how the resultant credit is shared in national carbon dioxide accounts.

What do we really need to do?

By focusing on Annex 1 countries (the big carbon polluters), the Kyoto Protocol goes to the heart of the matter. Its failing is that it is wholly inadequate; its targets are far too low, and too far in the future to make any significant difference. The Kyoto Protocol also excludes emissions from international transport – both planes and ships – which contribute significantly to the problem. Although there is growing pressure for a much stronger post-Kyoto deal, progress is slow. A wait-and-see attitude seems to be taking hold.

Some countries and cities have taken the lead in announcing their own targets. For example, the UK is currently putting in place legislation to meet a domestic target of a 60 per cent cut by 2050. The EU has agreed that by 2020, a fifth of energy must come from renewable sources and that countries must cut their emissions to 20 per cent below 1990 levels, and the European countries are challenging other countries to do the same. This is a wonderful symbolic gesture, but is it enough? George Monbiot suggests that to avoid a temperature rise of more than 2°C, and to meet an equitable emission quota, the UK needs to cut emissions by 90 per cent by 2030, a much tougher target. Other industrialized countries will need to do the same or more.

Solving global warming is not only a technical exercise and nor does the solution lie in reductionist accounting: we can't simply quantify how much we need to stop emitting and make equivalent cuts in emissions from our existing economies, although that can, and must, play a part in the transition.

Really, what is required is a fundamental shift in our civilization, in how we live. Global warming is like an alarm bell alerting us to some things that are very, very wrong with the world's economy and how we interact with each other. Solving

Section Three

Getting Personal

Chapter Ten

What We Can Do

*This chapter presents a summary of strategies to combat global
warming, and presents them in a form that individuals can
implement. Our contribution does count! We can reduce our
personal carbon footprint, join existing climate action groups,
lobby councillors and parliaments, and educate friends, family,
and neighbours. And get your money out of high-risk
seaside investments.*

Read on, and begin to change the world. We can do this in the
comfort of our own homes, and also by examining more
critically the ways in which we get around, the things we buy,
and the things we throw away. We can also become global
warming activists and help others to see the errors of their
ways.

Be energy-wise

Being energy-wise begins at home and in the workplace, and the starting point is efficiency. We may not be able to do without electricity from a coal-fired power station, but we can, at least, begin to use less of it, and save some money at the same time. Next, there are things we can do that may cost a bit up front, but will probably save us money in the long run. And we get to save the planet, too.

Efficiency in the home and at the office

A lot of electricity goes into heating and/or cooling our homes. Depending on the season and where in the world we live we can reduce our heating and cooling load by turning the thermostat or air conditioner down a few notches. An average European household will save about a tonne of carbon dioxide a year by reducing the thermostat by just 1.5°C. Put on or take off a layer or two of clothing instead.

If you live in a cold climate, don't heat the house while you are away for the weekend, and don't heat unused rooms. Buy a feather duvet and turn the heating down at night. Stop being a wimp: if your great-grandmother survived without central heating, so can you.

If you depend on central heating or cooling, get the system cleaned and serviced every few years to keep it working at maximum efficiency.

There's nothing like a hot bath in winter, but you could still enjoy one with the hot-water thermostat cooler by a degree or two; experiment with what works. If you use hot water only during certain parts of the day – in the evenings, say – install a timer switch which can be set to turn the power on and off at preset times.

And while the plumber is there, ask him or her to make sure your hot-water pipes and cylinder are well insulated.

Some of us have become reliant on energy-intensive appliances like dishwashers and washing machines, and older models are less likely to be as energy-efficient as newer versions. This by itself is not a good reason to rush out and buy a new one. Just use the old one as sparingly and efficiently as possible. Don't run half-loads; use cold- rather than hot-wash cycles; and use as little detergent as possible. While that may not keep global warming at bay, it will place less stress on the environment.

If you use a computer, turn it off when you're not using it. If you are just taking a coffee break, at least turn off the monitor which draws the most energy. Some models can be set up to "go to sleep" if, after a set time period, there has been no mouse or keyboard activity.

And when you take that coffee break, don't fill the kettle with more water than you need.

Most small appliances revert to standby mode instead of switching off completely. So turn appliances off at the wall and write letters to the manufacturers – and send them a complimentary copy of this book.

Save water. It takes a lot of energy to get the water to your taps.

Making changes
Energy-saving light bulbs use up to 80 per cent less energy and last up to ten times longer than conventional lightbulbs, and they are not even that much more expensive. Use them. And switch them off when you are out of the room.

Homes in cold climates lose most of their heat through their windows. Install double-glazing, which also cuts down on noise from outside, and get draught-excluders fitted on your external doors. If this is too expensive, keeping shutters, blinds and curtains closed at night will help to conserve heat loss, or, in hot, sunny climates, it will keep rooms cool during the day.

How well is your home insulated? The place to start is up in the attic, and the insulation materials required are relatively cheap and easy to install. This will keep your home warmer in cold weather, and cooler in the heat.

Most countries use coal to generate electricity. It is wonderfully convenient, but grossly inefficient, and each time you flip a switch a few more lumps of coal get burned and a few more kilograms of carbon dioxide go up the chimney. Where possible, wean yourself off electricity. Cook and heat with natural gas. It's more efficient and less carbon intensive.

If you live anywhere where the sun shines, a solar water heater is a must. Prices depend on how complex the system is and how much plumbing needs to be done. When the sun doesn't shine your conventional hot-water cylinder kicks in, so you're always assured of hot water.

If you are totally hooked on electricity, check whether one of the utilities servicing your area offer "green", or renewable, power. But investigate first – not everyone defines "green" power in the same way, and it may be simply a marketing ploy.

If you are in the market for a new electrical appliance, do some research first. Get one that has a good energy-efficiency rating, or see if it has been tested by an independent reviewer. Some may even carry a "green" or energy-efficient label. Is it for real? Ask the salesperson for details and make them work for their commission. Lobby your government to regulate around this and provide you with the information you need.

Getting around

Apart from walking and using a bike, there are not too many modes of transport that don't pump loads of carbon dioxide into the air. About 12 per cent of our carbon dioxide emissions come from cars, planes, and ships. So use them less frequently, and use them wisely.

Use public transport wherever you can (if neither walking nor biking is an option).

If you can't commute on public transport, see if you can form a lift-club or car-pool with your colleagues.

If you have a car, keep it serviced and check the tyre pressure regularly to help it run more efficiently. And the more conservatively you drive, the less fuel you use – drive as if there were an egg under your accelerator foot.

If you own a 4×4 or pickup truck but not a farm, hang your head in shame.

Don't carry lumpy, non-streamlined things on your roof-racks if you can help it, and don't take so much luggage with you; the heavier and more wind-resistant your car, the more fuel you will use.

Plan local holidays, rather than travelling long distances.

Finally, and most importantly, just travel less. Do we really need to fly to that business meeting? Wouldn't a video- or tele-conference be just as effective? Air travel is a major contributor to global warming.

Your daily bread
The things we consume and the things we throw away all play their part in global warming. Think about where the product you want was made, how it was made, and whether you really need it. And before you throw it or its packaging away, consider if you could reuse or recycle it. Repeat after me: "Refuse (as in, say no), reduce, reuse, recycle."

Buy local. Food that has had to travel comes with invisible carbon "food miles". Do we really need out-of-season apples from the other side of the world?

Organic food uses fewer or no fertilizers and poisons. Most fertilizer manufacturing is energy intensive, so avoid fertilizers. Poisons kill the bad bugs as well as the good ones which help to keep carbon in the soil, where it belongs.

Plastics are made from fossil fuel carbon compounds. Carbon is released years later when these plastics degrade. Check that the packaging you buy is recyclable, and then make sure you dispose of it at a recycling centre. Even better, send excessive packaging back to the producer or retailer to help them get the message.

Landfill sites generate massive amounts of methane, a green-house gas twenty-three times more powerful than carbon dioxide. The less rubbish you send to the dump, the less methane gets released.

Any cash left over? You may consider looking for a climate-friendly investment opportunity – like stocks and shares in a renewable energy producer. Most stock exchanges offer socially responsible funds. See if any of them include environmental or climate-friendly investment criteria. You could buy a solar water heater for your neighbour or pen-pal across the world.

Become a climate activist
Write a letter to your local newspaper the next time you read an article about global warming. You can quote us!

Write to your political representative. Ask why he or she is doing so little about global warming and make a suggestion.

Need more information? Surf the internet – it's full of information, though it is also the home of contrarians and climate change-deniers, so surf wisely.

Most importantly, as a serious activist, you will have to learn to walk the talk. There are a few good websites that allow you to calculate your own personal carbon footprint – make it smaller.

Select Bibliogaphy

Dow, Kirstin, and Downing, Thomas E., *The Atlas of Climate Change: Mapping the World's Greatest Challenge,* Earthscan, 2006

Flannery, Tim, *The Weather Makers*, Atlantic Monthly Press, 2005

Gupta, Joyeeta, *Our Simmering Planet*, 2001

Joubert, Leonie, *Scorched*, Wits University Press, 2006

Kolbert, Elizabeth, *Field Notes from a Catastrophe: Man, Nature and Climate Change*, Bloomsbury, 2006

Lomberg, Bjørn, *Skeptical Environmentalist*, Cambridge University Press, 2001

Monbiot, George, *Heat*, Allen Lane, 2006

Stevens, William K., *The Change in the Weather*, Dell Publishing, 1999

Recommended Websites

To see emission trends from countries and sectors.
http://earthtrends.wri.org/

For up-to-date info on the flex mechs (especially CDM).
http://www.uneprisoe.org/

UN Framework Convention on Climate Change.
http://unfccc.int

For how to argue with a climate change-denier.
http://gristmill.grist.org/skeptics

For tips on energy use.
http://www.rmi.org/

Climate action network.
www.climatenetwork.org

Intergovernmental Panel on Climate Change.
www.ipcc.ch

For real science from climate scientists.
www.realclimate.org

To calculate your own emissions.
www.carbontrust.co.uk
www.safeclimate.net/calculator
www.myclimate.org